Excimer
Laser
Lithography

Excimer Laser Lithography

Kanti Jain

Anvik Corporation

SPIE OPTICAL ENGINEERING PRESS

A Publication of SPIE–The International Society for Optical Engineering
Bellingham, Washington USA

Detler H Finzauer

Library of Congress Cataloging-in-Publication Data

Jain, Kanti, 1948-
    Excimer laser lithography / Kanti Jain.
        p.   cm.
    Includes bibliographical references.
    ISBN 0-8194-0272-9 — ISBN 0-8194-0271-0 (soft).
    1. Excimer lasers—Industrial applications. 2. Microlithography.
I. Title.
TA1695.J35 1990   89-24240
621.381′531—dc20

ISBN 0-8194-0272-9 (hardbound)
ISBN 0-8194-0271-0 (softbound)

Published by
SPIE–The International Society for Optical Engineering
P.O. Box 10
Bellingham, Washington 98227-0010

The text of this book was prepared camera-ready by the author.
Book design: Kelly A. Anderson

Printed in the United States of America

Second Printing

To My Parents

Contents

Preface

This book provides an overview of the rapidly emerging technology of excimer laser lithography. It is designed both for professionals actively involved in semiconductor lithography as well as for others who desire to obtain a state-of-the-art familiarity with the issues and developments that will guide optical lithography to the sub-half-micron regime. The book has evolved from a short course the author has given in conjunction with several international conferences in the past year. Expecting that the course will attract, as it has, attendees from a diverse set of disciplines, including lithography, lasers, optics, semiconductor manufacturing equipment, wafer fabrication processes, resist technology, and materials processing, special emphasis has been put on providing as broad a coverage as possible of the field of excimer laser lithography; this breadth is reflected in the scope of the book.

The organization of the book and the major topics it addresses are as follows. The introduction presents a discussion of the motivation for short-wavelength lithography in general and excimer laser lithography in the deep ultraviolet wavelength region in particular. The case for microlithography with excimer lasers is expanded in the next chapter in which a comprehensive review of the various available sources of ultraviolet radiation, both coherent and incoherent, is presented. The third chapter briefly summarizes a variety of attempts made by researchers to achieve fine-line lithographic patterning using conventional, non-excimer laser sources. The reader, thus having been familiarized with a thorough background, then proceeds to the largest section in the book, namely, microlithography with excimer lasers. Beginning with the first submicron exposures, this chapter presents various excimer laser exposure tool concepts and system requirements. Developments reported in excimer laser projection printing on a variety of commercial lithographic machines are reviewed. Recent results obtained with full-field scanning projection systems as well as step-and-repeat tools are examined. Several related aspects of the complete excimer laser microlithography system, such as alignment techniques, illumination concepts, and optical component needs, are discussed. This is followed by a chapter on excimer laser sources, which describes the fundamentals as well as various operational parameters of excimer lasers. Excimer laser performance issues are discussed from two points of view: availability from laser manufacturers and requirements for various practical lithographic systems. Next, an overview of developments in resist technology for excimer laser lithography is provided. Here, investigations of a number of resist materials exposed with different excimer laser wavelengths are summarized, efforts currently in progress in various research and development programs are discussed, and key issues concerning photoresist technology as it relates to excimer laser lithography are examined. The next chapter deals with applications of excimer lasers in processes other than conventional lithography, such as etching and deposition. Finally, in the Outlook section, the current status of the practical implementation of excimer laser lithography in the semiconductor industry is reviewed and future directions in optical lithography are examined in view of these advances.

The author would like to thank all his colleagues whose work he has been able to use in the book. He expresses his appreciation for the enthusiasm shown by all the professionals who have taken the course at numerous conferences. Appreciation is also due to SPIE for extending the invitation to write the book and for being most helpful and patient during its preparation. Finally, it is the author's pleasure to thank his wife Vijaya and his sons Vivek and Anshul, whose support and understanding made it all possible.

Briarcliff Manor, New York   *Kanti Jain*
September 1989

1. Introduction

1.1. PROGRESS IN MICROLITHOGRAPHY

The ever-increasing requirements on the device density and performance of semiconductor chips have motivated extensive research in various lithography technologies in recent years. Since the beginning of the integrated-circuit (IC) era, the dimensions of the individual devices on the ICs have shrunk steadily and rapidly, along with improved device design concepts and increased chip sizes. Such consistent progress in IC technology has been made possible most significantly by advances in microlithography tools. In Fig. 1.1 we trace the semiconductor industry's march toward ever-decreasing minimum feature sizes in device fabrication over the last two decades and present a projection for its continuation in the next decade. Beginning with the late 1960s when the industry was producing 1-kilobit (Kb) memories, the figure shows how the bit density on memory chips has quadrupled approximately every 3 to 3 1/2 years, reaching volume production of 4-megabit (Mb) chips toward the end of the 1980s. Note that accompanying this progress in bit density has been a phenomenal decrease - by greater than an order of magnitude, from ~ 10 microns for the 1-Kb chips to 0.7 micron for the 4-Mb chips - in the minimum linewidths. Continuation of this trend is certain: developmental 16-Mb devices with 0.5-micron geometries have already begun to appear, research efforts toward the 256-Mb era are in progress, and it is likely that toward the end of this century we shall see 1-gigabit (Gb) memory chips with minimum linewidths under 0.2 micron.

1.2. LITHOGRAPHY EQUIPMENT TECHNOLOGIES

In Fig. 1.1 we have also indicated the type of lithography equipment technology most commonly employed in each device generation. Notice that the primary manufacturing tools to date have all been optical: from the early contact and proximity tools to the full-wafer

1

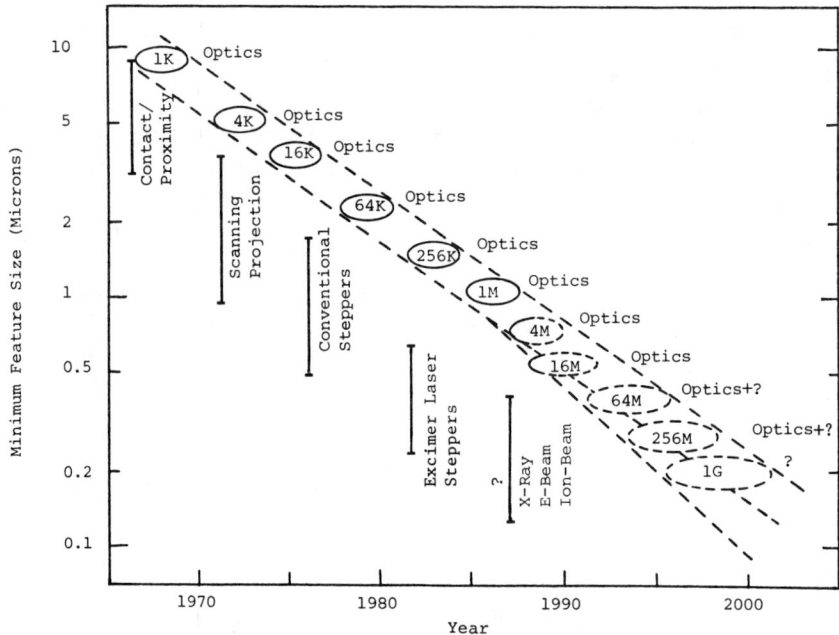

Fig. 1.1 Progress in microlithography, depicted as decreasing minimum feature size, over the last two decades and projections for the next decade. Also shown is the corresponding increase in integrated-circuit-chip bit densities. The bit density refers to that of the leading dynamic random-access memory chip. Various lithography tools and their ranges of applicability are also indicated.

scanning projection systems of the 1970s to the step-and-repeat reduction machines that have dominated the 1980s. Although there has been considerable progress in research in alternate lithography technologies, including electron-beam, x-ray and ion-beam, they have not been able to displace optical lithography on the manufacturing floor in any significant way. Although numerous electron-beam lithography systems may be found in semiconductor fabrication facilities around the world, their use has been confined to mask making and certain specialized, low-volume applications on wafers. The historical dominance of optical lithography tools is expected to continue in the next decade. The reasons for this assertion are straightforward and easy to understand. First, optical tools remain the most economical way to print patterns on wafers. Second, since optical lithography is the most entrenched wafer exposure technology in the industry, semiconductor manufacturing lines employing optical tools incur the least cost in upgrading their tools as the industry moves from one device generation to the next. Finally, due to the emergence of excimer laser lithography, optical tools will continue to deliver the ability to print finer and finer geometries on wafers at high throughput rates, thus extending their dominance into the half- and sub-half-micron regimes, which had been previously thought to be out of reach for optics. For wafer lithography, it now appears certain that the two major candidates that have been suggested as alternate technologies, x-ray and electron-beam, will not be strong contenders until perhaps the 256-Mb, or possibly the 1-Gb, device generation: electron-beam due to its uneconomical throughput, and x-ray due to problems on all three fronts - cost effectiveness, timeliness, and freedom from technical difficulties.

1.3. DEEP ULTRAVIOLET LITHOGRAPHY

In optical lithography, a major thrust in the last several years has been a push toward lithography in the deep ultraviolet (DUV) (200-300 nm) spectral region. This is due to the numerous advantages that imaging

with shorter illumination wavelengths offers. The most
well-known of these advantages is the improvement in
resolution, which may at once be observed from the
expressions for the smallest resolvable feature in the
cases of projection printing and proximity printing, as
given by, respectively,

$$w = \lambda/(2NA) \qquad\qquad (1)$$

and $$w^2 = \lambda z/2, \qquad\qquad (2)$$

where w is the resolution, NA is the numerical aperture
of the imaging system in the projection printing case,
and z is the mask-to-wafer gap tolerance in the case of
proximity printing. Further, if one looks at the depth
of field (DOF), given by

$$DOF = \lambda/NA^2, \qquad\qquad (3)$$

one observes from Eq. (1) that for a given resolution
w, the use of a shorter exposure wavelength permits a
smaller NA, and therefore produces a larger depth of
field. By substituting for NA from Eq. (1) in Eq. (3),
this argument may be expressed as

$$DOF = 4w^2/\lambda. \qquad\qquad (4)$$

In the case of proximity printing, a shorter wavelength
allows one to use a larger value of z. Finally, the
images produced by a projection exposure system using a
shorter wavelength have steeper sidewall profiles. This
may be understood with the help of what is known as the
optical modulation transfer function (MTF), illustrated
in Fig. 1.2(a). The MTF in a high-resolution projection
lithography system is a measure of the fidelity with
which the total microlithographic process reproduces a
perfect square-wave object pattern as an image in the
resist medium. Figure 1.2(b) describes the parametric
dependence of the MTF-vs-spatial frequency curve on the
exposure wavelength, and shows that for a given spatial
frequency, i.e., for a given image size, the MTF
increases with decreasing wavelength, which directly
translates into steeper image profiles.

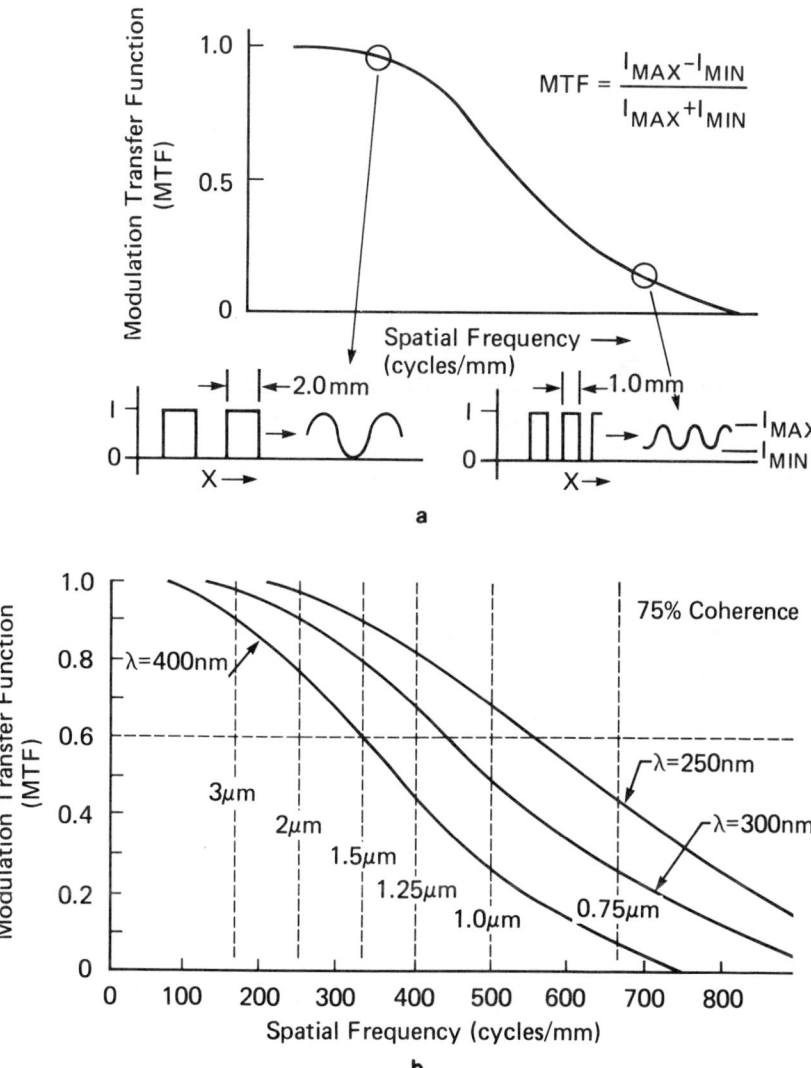

Fig. 1.2 (a) Definition of optical modulation transfer function and (b) its dependence on the exposure wavelength.

1.4. EXCIMER LASER LITHOGRAPHY

The above considerations have promoted a careful assessment of various sources of DUV radiation for microlithography. These sources, both coherent and incoherent, will be described in detail in the next chapter. However, we shall quickly summarize some key points regarding DUV sources. The extendibility of the currently used high-pressure mercury arc lamps for DUV lithography is severely limited, not only due to their very poor brightness in the desired 250-nm spectral region, but also due to the fact that they generate excessive heat and distortions as a result of the unwanted long-wavelength radiation. The application of 'conventional' ultraviolet lasers (e.g., argon-ion) for high-resolution imaging has long been thought to be impractical due to the interference patterns called speckle that laser sources routinely produce. In contrast, excimer lasers are ideal as light sources for DUV lithography for a number of reasons. Their high power output provides extremely short wafer exposure times, and their spatial incoherence - unlike other lasers - rids the exposure of speckle. Further, as a result of the availability of greater exposure power, excimer lasers make the task of developing suitable photoresist materials simpler. Optically, excimer lasers may be made to operate with spectral linewidths as narrow as a few picometers, making it possible to design high-performance all-quartz stepper lenses.

As a consequence, most of the recent effort in DUV lithography technology has focused on the use of excimer lasers. The results obtained with excimer laser lithography have been so promising, and progress in its practical implementation in manufacturing tools so rapid, that it now appears likely that excimer laser lithography tools will become the mainstay of half- and sub-half-micron wafer fabrication lines in the next decade. Since the first reports by Jain et al. [1-3] on the proposal and demonstration of fast, speckle-free, high-resolution excimer laser lithography, the interest in this technology has grown widely [4-76]. Reports published in the open literature numbered more than 75 by the middle of 1989, and include activities at several establishments in the U.S. [1-3,5,6,8-19,23, 26,27,29,35,36,38-42,48,50-54,63-67], Japan [4,20,37, 43,45-47,49,55-58,61,62,68-74], and Europe [7,21,22, 24,25,28,30-34,44,59,60,75,76]. Table 1.1 provides a list of the several industrial as well as academic organizations active in various aspects of excimer laser lithography. Many of these developments are discussed in the following chapters.

Excimer Laser Lithography

Table 1.1. Organizations active in excimer laser lithography.

U.S.	JAPAN	EUROPE
IBM	Nikon	Karl-Suss (W. Germany)
AT&T	Canon	ASM (The Netherlands)
Texas Instruments	Toshiba	Rutherford Lab. (UK)
Hewlett-Packard	Matsushita	Acad. Sci. (USSR)
Image Micro Systems	NEC	Zeiss (W. Germany)
Shipley	Hitachi	
GCA	Fujitsu	
Perkin-Elmer	Oki	
Ultratech	NTT	
MIT	Sony	
Stanford	Mitsui	
U. Calif.-Berkeley		

2. Sources for Deep Ultraviolet Lithography

The wide variety of sources that emit ultraviolet radiation can be broadly divided into two groups: lamp, or incoherent, sources and 'laser-like,' or coherent, sources. In the latter class, we include all techniques and devices for UV light generation that have their basis in stimulated emission of radiation. The lamp sources may be further subdivided into high-pressure arc lamps, deuterium lamps, etc. In this chapter we discuss the properties of the various subclasses of both lamp and laser-like sources and show why, among these, excimer lasers are the most promising for short-wavelength optical lithography.

2.1. LAMP SOURCES

2.1.1. High-Pressure Arc Lamps

These sources are usually mercury or mercury-xenon arc lamps with electrical input power in the 0.5-1.5 kW range. The relative partial pressures of Hg and Xe as well as the total pressure of the discharge mixture determine the spectral distribution of the light output among the deep UV, mid-UV ($\sim$ 300-350 nm), and near or conventional UV ($\sim$ 400 nm) regions. Since a substantial fraction of the input power is converted into heat, it is not considered practical to use lamps of this type that would be much more powerful than those currently available because of the limitations imposed by the thermal instability of the quartz envelope. Their most attractive feature is the small, well-defined source size, usually in the shape of a point or a line, which makes them optically very convenient to handle. This, in combination with their adequate power output in the conventional UV region, has made them the most commonly used light sources for optical lithography. However, their extremely poor efficiency at shorter wavelengths has severely limited their usefulness in the deep UV region, especially as light sources in half- and sub-half-micron reduction steppers.

9

2.1.2. Deuterium Lamps

Deuterium lamps have a broadband emission in the
deep UV region and are found useful in situations where
blanket exposures in a wide wavelength spectrum are
desirable. However, with typical electrical inputs of
200 W, their deep UV efficiency is too low (<0.01%) for
them to be considered practical for microlithographic
exposure systems. Attempts to increase their light
output have been unsuccessful due to various problems
associated with out-diffusion of the deuterium gas,
plasma instability, and a short lifetime.

2.1.3. Microwave-Excited Lamps

These are mercury discharge lamps, but different
from the high-pressure arc lamps of Sec. 2.1.1 in the
sense that the discharge is excited at a microwave
frequency, rather than dc or low-frequency ac. The
microwave excitation causes the discharge plasma to be
confined within close proximity of the envelope walls,
thus reducing reabsorption of the emitted light by the
discharge gas. They have a wide-band emission, and the
deep UV efficiency integrated over this band is high.
Their main drawback is the extremely large source size
(commonly a 2-3 cm diameter sphere or a 20-30 cm long
cylinder), which renders them inconvenient from the
optical designer's point of view. They are attractive,
however, for blanket exposure applications, e.g., for
processes employing multilayer resist systems.

2.2. LASER-LIKE SOURCES

In this group we include those light sources that
generate the ultraviolet photon by stimulated emission
of radiation. Laser-like sources may also be termed
coherent sources but, as we discuss below, in a broad
sense, it is more appropriate to call them only
temporally coherent. Among these sources, we make a
further distinction between 'primary' and 'secondary'
laser-like sources. Primary sources are themselves
lasers, i.e., devices in which the UV photon is emitted

by a fundamental transition in a lasing species. This fundamental transition in the lasing atom or molecule may be, for example, between electronic, vibrational, or rotational levels. Secondary sources are those that are produced by shifting the wavelength of a primary laser source using one or more of various frequency conversion methods such as harmonic generation, sum-frequency mixing, and stimulated Raman shifting. Such frequency conversion takes place through a nonlinear interaction in an optical medium either between two photons or between a photon and one or more elementary excitations in the medium.

2.2.1. Primary Laser-like Sources

A large number of atomic and molecular species have been successfully investigated for laser action in a wide range of UV wavelengths. Among these systems, the most powerful are the excimer lasers, which were first developed in the mid-1970s. Prior to the advent of these highly efficient systems, UV lasers operating in the 190-360 nm region had been limited to multiply ionized noble gas lasers, singly ionized metal-vapor lasers, and nitrogen lasers. In this section we present a summary of the key characteristics and features of each of the above types of primary sources of coherent UV radiation.

2.2.1.1. Multiply Ionized Rare Gas Lasers

The lasing medium in this class of lasers is a multiply ionized state of a noble gas atom. Species in which UV laser action has been observed include the ions NeIII, NeIV, and NeV, ArIII and ArIV, KrIII and KrIV, and XeIII and XeIV [77,78]. As in the case of the more well-known singly ionized visible rare gas lasers, lasing in these multiply ionized species is produced by exciting a low-pressure (typically on the order of a few tens of millitorrs) longitudinal high-voltage arc discharge. However, since the latter have lower gain, they require higher pumping threshold current densities - typically 10,000 A/cm^2. Further, they also have poor efficiencies (<0.01%), and obtainable output average powers range between 10^0 to 10^3 mW. Most of the laser transitions in this class are pulsed; however, some, in the vicinity of 350 nm, can be made to run in the continuous-wave (cw) mode. Some of the wavelengths belonging to this type of lasers are shown in Fig. 2.1.

2.2.1.2. Singly Ionized Metal Vapor Lasers

The singly ionized states of certain metal atoms have transitions in the UV that can be made to lase under appropriate excitation conditions. Most suitably, a hollow-cathode electric glow discharge has been effectively used to produce and excite a variety of metal vapors, including copper, silver, gold, and cadmium [79,80]. The hollow cathode is made of the metal to be excited, which is produced in vapor form by sputtering of the cathode when a discharge is initiated in a rare gas buffer at a pressure of a few torrs. These metal-ion lasers can be made physically small and run in the cw mode, but the best overall efficiencies achieved to date (<0.002%) are even worse than those for multiply ionized noble gas lasers. Although laser transitions at a number of wavelengths have been obtained, including the shortest wavelength - 224 nm in AgII - for a cw laser, no laser of this type has been able to produce an average output power greater than a few milliwatts in the deep UV. Some representative metal vapor laser lines are indicated in Fig. 2.1.

2.2.1.3. Excimer Lasers

A detailed discussion of excimer lasers is presented in Chapter 5. Here, we summarize some of their key features. Excimer lasers are the most powerful of all UV laser sources (see Fig. 2.1 for a clear comparative picture). Their operation derives from the population inversion that can be created in the electronic states of a metastable rare gas halide molecule when a high-pressure (>1 atm) mixture of the rare gas and the halogen (or halogen-bearing compound) is excited in a high-voltage pulsed discharge [81,82]. Properly designed excimer lasers can have extremely high efficiencies, can achieve very high peak

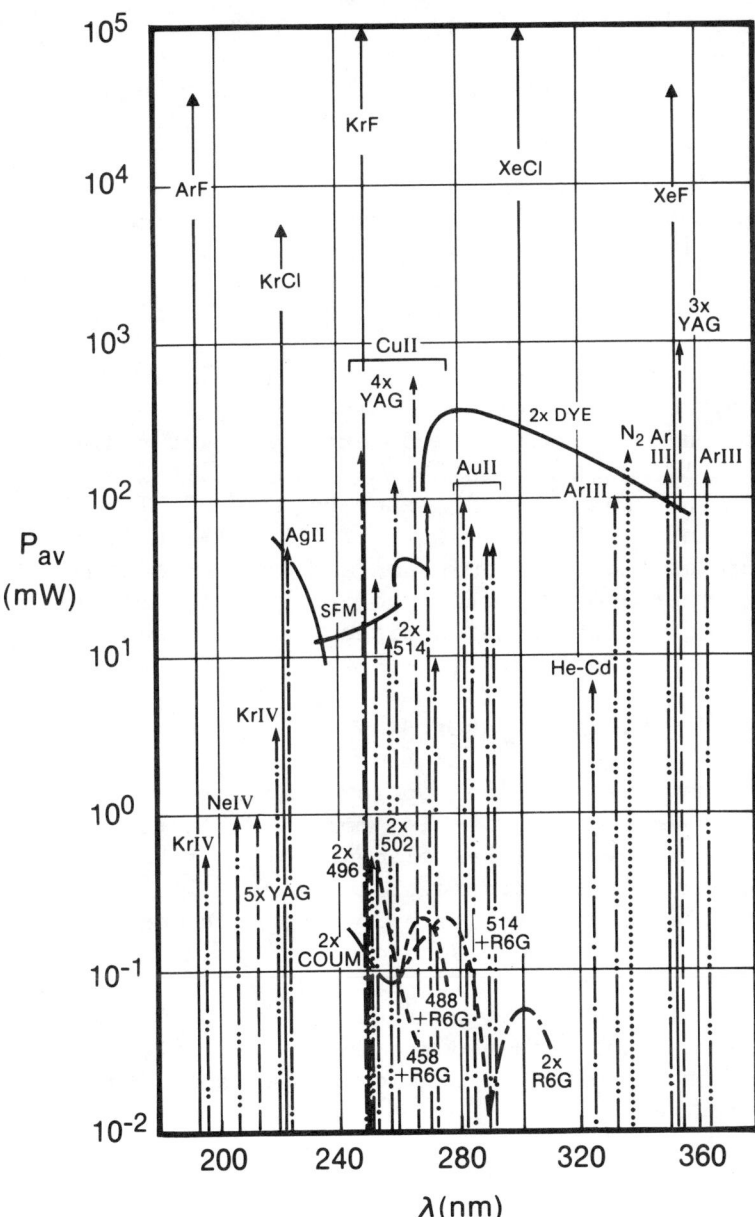

Fig. 2.1   Comparison of the average output powers
available from various primary and secondary sources
of coherent ultraviolet radiation.

and average power outputs, and can be made compact.
Wavelengths available from excimer lasers range from
350 nm to below 150 nm. In addition to the high power,
an important property of excimer lasers that sets them
apart from other, 'conventional' lasers is the lack of
good spatial coherence in their beams. When an object
is illuminated with a spatially coherent wavefront, any
scattering at an optical surface will cause different
parts of the wavefront to interfere constructively and
destructively in the sample plane and produce a random
pattern called speckle. This problem has historically
hindered the application of lasers in high-resolution
lithography. However, illumination by an excimer laser
produces no speckle because excimer laser radiation has
extremely poor spatial coherence. This is due to the
fact that an excimer laser beam is highly multimode,
which in turn is a result of the almost superradiant
nature of the laser emission and the large mode volume.
The coherence properties of excimer lasers and their
comparison with other sources is discussed at greater
length in Chapter 5.

high # of spatial modes

2.2.1.4. Nitrogen Laser

The nitrogen laser operates at a wavelength of 337
nm. Its lasing species is molecular nitrogen, and the
lasing transition takes place from a rotational level
of an excited electronic state of the nitrogen molecule
to a rotational level of a lower electronic state [83].
Nitrogen lasers use discharge conditions and geometries
similar to those used for excimer lasers, but have poor
efficiencies and low average output powers. Typical
pulse energies obtained are in the few millijoule range
- two orders of magnitude lower than those from excimer
lasers.

2.2.2. Secondary Laser-like Sources

As mentioned, secondary sources of coherent
radiation are those produced by shifting the wavelength
of a primary coherent source by frequency conversion
through optical nonlinear interaction in a suitable
medium. Although most of the known secondary means of

generating coherent deep UV radiation require very large pump powers, have poor efficiencies, and are therefore unattractive as lithography exposure sources, some do present certain attractive features. In the following we give brief descriptions of the three most well-known and versatile frequency conversion methods.

2.2.2.1. Harmonic Generation

This is a method in which the passage of a high-power laser beam of an optical frequency f through a nonlinear optical medium results in the generation of a certain harmonic of f. Depending on the nature of the nonlinearity of the medium and various physical as well as geometrical parameters, the generated harmonic may be at a frequency 2f, 3f, 4f, etc. (second harmonic, third harmonic, etc.) [84,85]. From the fundamental quantum mechanical viewpoint, the generation of the n-th harmonic of an input frequency f is the creation of a photon at a frequency nf when n photons each of frequency f combine through an interaction in a medium. Since the n input photons must be combined coherently and since the interaction must preserve coherence in the harmonic beam, certain 'phase-matching' conditions must often be met. These conditions impose well-defined orientational constraints on various physical aspects of the optical arrangement in the frequency conversion system.

A large variety of materials have been used as nonlinear media for harmonic generation; these include crystalline materials, various liquids, and several gases and vapors. Most common among these media are optically nonlinear and birefringent crystals such as ADP (ammonium dihydrogen phosphate, $NH_4H_2PO_4$) and KDP (potassium dihydrogen phosphate, KH_2PO_4). Due to the phase-matching requirements mentioned above, all such crystals must be oriented and cut along certain well-defined crystallographic directions and, further, due to the dependence of phase matching on the refractive indices, the crystals must often be maintained at a controlled temperature.

An example of a commonly used frequency-doubled system is the Ar-ion laser, the 514-nm line from which is doubled in an ADP crystal to provide coherent cw radiation at 257 nm. Another well-known system is the pulsed Nd:YAG laser at 1064 nm, which, using KDP as the nonlinear conversion medium, is frequency-tripled to generate coherent emission at 355 nm, and frequency-quadrupled to provide a secondary source of DUV laser-like radiation at 266 nm. Both of the above systems are commercially available. In addition to the generation of harmonics of discrete laser lines, one may just as easily frequency-double a dye laser, in which case a source of tunable coherent UV radiation is produced.

2.2.2.2. Frequency Mixing

The basic principle of the technique of frequency mixing (or, sum-frequency mixing, to be more specific) is similar to second-harmonic generation in that the energies of two input photons are added to produce an output photon with an energy equal to the sum of the energies of the input photons. The difference is that in second-harmonic generation the two photons are from the same source and are identical, whereas in frequency mixing they are usually from different sources and may have different frequencies. Thus, it is possible to mix the output of an Ar-ion laser with that of a cw dye laser, producing cw tunable UV. Another good example is the mixing of an excimer laser beam with a pulsed dye laser beam, producing an intense and coherent source of pulsed and tunable UV radiation. Some of the secondary sources of DUV radiation that belong to this category are shown in Fig. 2.1.

2.2.2.3. Stimulated Raman Shifting

Spontaneous Raman scattering is the process by which radiation interacting with a material experiences a frequency shift as a result of inelastic scattering of the incident photons from certain characteristic excitations (e.g., molecular vibrations, electronic transitions, etc.) of the medium. When the input power is made large enough, this frequency-shifted scattered radiation can be stimulated. Whereas the conversion efficiency in the spontaneous Raman effect is typically $\sim 10^{-7}$, it can be as large as 0.7 in the stimulated case. Thus, intense coherent light can be produced at different wavelengths by stimulated Raman scattering. The output frequencies can be both down-shifted (Stokes components) and up-shifted (anti-Stokes components) from the incident laser frequency by multiples of the frequency of a molecular vibration (or other elementary excitation) characteristic of the Raman medium. Since extremely high peak powers are available from excimer lasers, they are ideally suited for stimulated Raman shifting. Each of the primary excimer laser wavelengths can be shifted to several new wavelengths by selecting different Raman media. For example, one may obtain many different sets of Stokes- and anti-Stokes-shifted lines spanning the range from 190 to 415 nm by stimulated Raman scattering in H_2 , D_2 , CH_4 , and N_2 using ArF, KrF, KrCl, and XeCl lasers [86]. Further, the distribution of the input pulse energy into various components may be tailored by adjusting the pressure in the Raman cell and/or the optical parameters.

3. Lithography with Conventional Lasers

In this chapter we discuss the use of conventional UV lasers in microlithography. The term 'conventional' is meant to apply to those lasers whose output beams are highly coherent, both temporally and spatially. Such is the case for those laser systems in which the stimulated emission of the laser radiation is built up in a conventional two-mirror optical resonator cavity. As indicated previously in Sec. 2.2.1.3, the high degree of coherence - especially spatial coherence - of conventional lasers has historically rendered them unattractive for use in high-resolution lithography due to the problem of speckle. In illumination of an object with a spatially coherent wavefront, any scattering at an optical surface causes different segments of the wavefront to interfere constructively and destructively at the sample surface and generate a random pattern called speckle. This problem has limited the ways in which coherent lasers can be used for creating fine-line patterns. In this chapter we discuss these limitations and review various examples of patterning with coherent lasers.

One situation in which pattern delineation with conventional lasers is free from speckle is that in which one does not use a mask and in which there is no imaging of an optical pattern from an object plane into an image plane. Instead, the laser beam is focused onto the photoresist-coated substrate to the desired spot size and, as shown in Fig. 3.1, the spot is translated relative to the substrate, or vice versa, to expose the desired pattern in the substrate. This scanning-spot technique is, thus, a direct-write method, similar to electron-beam lithography. Such a system for patterning ceramic substrates for hybrid integrated circuits was developed by Takaba et al. at Nippon Electric Company [87]. This system used the 476-nm line from an Ar-ion laser to expose patterns in Shipley/AZ 1350 resist with a minimum linewidth of 20 microns. A similar system built by Eocom Corporation for exposing 40-micron-wide wiring patterns in multilevel printed circuit boards

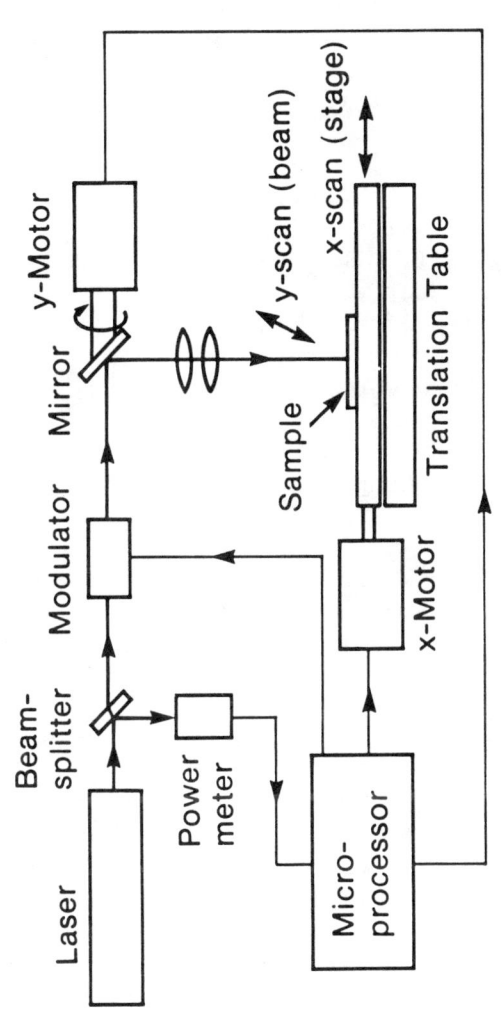

Fig. 3.1 A focused-laser scanning-spot lithographic exposure system. The desired pattern is generated by scanning the focused laser beam and the substrate relative to each other.

has been demonstrated by Westinghouse Electric Company
[88]. The 458-nm Ar-ion laser line has been used by
Becker et al. [89] in a focused-laser scanning-spot
lithography apparatus to expose <4-micron-wide lines
in a laboratory system. A focused-laser scanning system
with resolution in the vicinity of a micron has also
been reported [90]; this system uses a helium-cadmium
laser operating at 442 nm and large numerical aperture
optics to achieve its resolution.

In a more complex adaptation of the same scanning-
spot technique, ATEQ Corporation has developed a high-
throughput mask writer (see Fig. 3.2) using multiple-
beam scanning [91]. This system uses the 364-nm line
from an Ar-ion laser, which is split into eight parallel
beams by a beamsplitter. These beams are independently
modulated in an acousto-optic modulator and then swept
across the substrate by a rotating polygonal mirror.
Movement of the substrate by a translation stage at
right angle to the direction of beam-sweep and on-off
modulation of the beams exposes the desired pattern on
the substrate.

Although the scanning-spot techniques described
above successfully eliminate speckle, they suffer from
several major disadvantages. First, the scanning nature
of the exposure requires that the laser be operated
essentially in the cw mode. This has the undesirable
consequence of eliminating the most powerful of the
available laser sources, which all happen to be pulsed
(e.g., excimers and harmonics of Nd:YAG) with extremely
low duty cycles (<0.001%). Second, various technical
problems associated with building sophisticated two-
dimensional optical deflection systems become more and
more formidable as the resolution requirements become
tighter. And finally, by its very nature, the focused-
laser technique renders itself unusable in the majority
of the most common microlithography applications in
semiconductor device fabrication in which a large-area
patterned mask must be imaged on a wafer through highly
complex projection optics. Any attempt at imaging with
a conventional (i.e., coherent) laser results in the
reappearance of the speckle problem due to the spatial

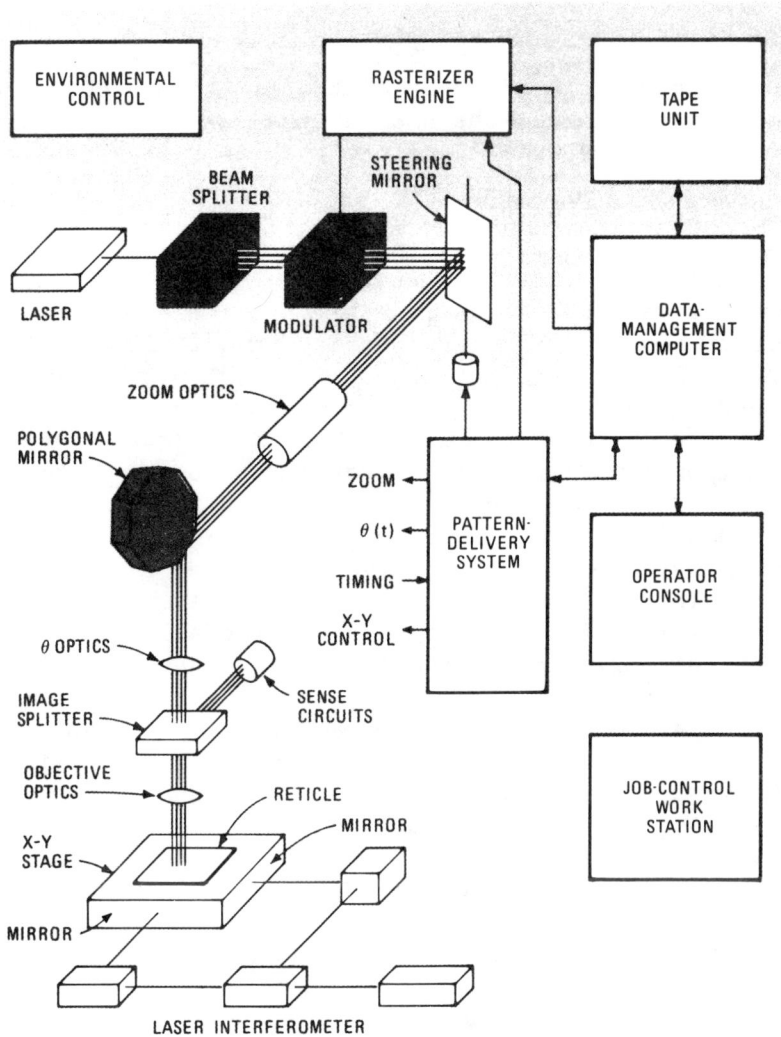

Fig. 3.2   Multiple scanning spot laser pattern
generator for mask-making, manufactured by ATEQ
Corporation. [From Ref. 91]

and temporal coherence of the source.

However, there are some techniques available that one may exploit to eliminate speckle in the exposure methods described above. Since the spatial coherence of a source directly depends on the emitting size of the source (the larger the source, the less coherent it is spatially), one may transform the laser beam, which is essentially a point source, into an extended source to reduce speckle. This method was used by Lacombat et al. [92], who scanned a 413-nm krypton laser beam in the entrance pupil of their projection system; by scanning areas of various sizes in this pupil plane, they were able to achieve different degrees of partial coherence. Since the transformation of a point source into an extended source is equivalent to using a large number of point sources, the reduction in speckle in the above situation may be thought of as a result of incoherent superposition of a large number of coherent wavefronts. This has been demonstrated by the experiments of Kozma and Christensen [93], who have shown (see Fig. 3.3) how the quality of the images of a set of bar targets made by exposures with an Ar-ion laser gradually improves as the number of superimposed coherent images increases.

Using a different approach to lithography with conventional lasers, Levenson [94] has demonstrated high-resolution image formation by conjugate wavefront generation in optically nonlinear media. In a nonlinear four-wave mixing interaction, illustrated in Fig. 3.4, two mutually counter-propagating pump waves interact with an object wave through the third-order nonlinear susceptibility of the conjugator medium. The image wave propagates in a direction exactly opposite to that of the object wave. Since the two pump waves are phase-conjugates of each other, the image wave is generated as the phase-conjugate of the object wave. Thus, if the object wave originates from a high-resolution mask, the image wave, in principle, will faithfully reproduce its intensity pattern. Optical systems employing such a process of image formation through conjugate wavefront generation automatically compensate for aberrations and distortions. Thus, theoretically, the method is capable

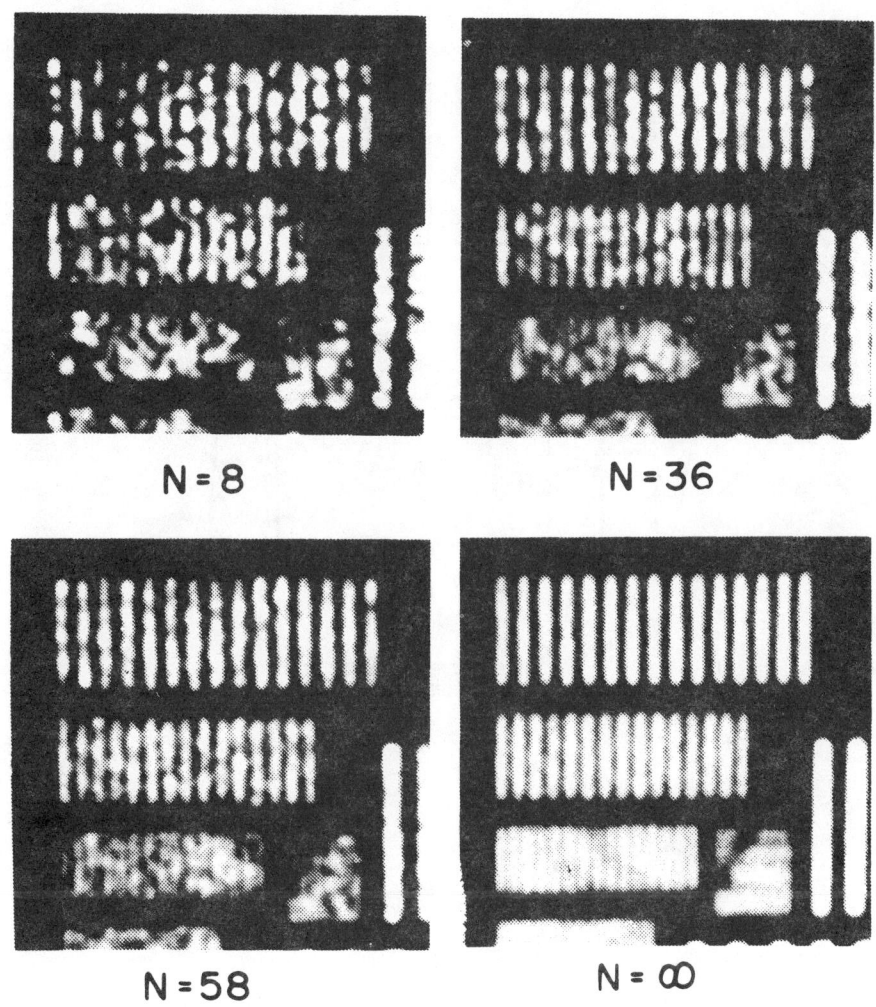

Fig. 3.3   Increase   in   bar   target resolution as the
number, N, of superimposed coherent images, each with
an independent speckle pattern,   increases. The   light
source was   a   514.5-nm   Ar-ion laser.   For the N = ∞
image, the illumination was made incoherent by using a
moving diffuser. [From Ref. 93]

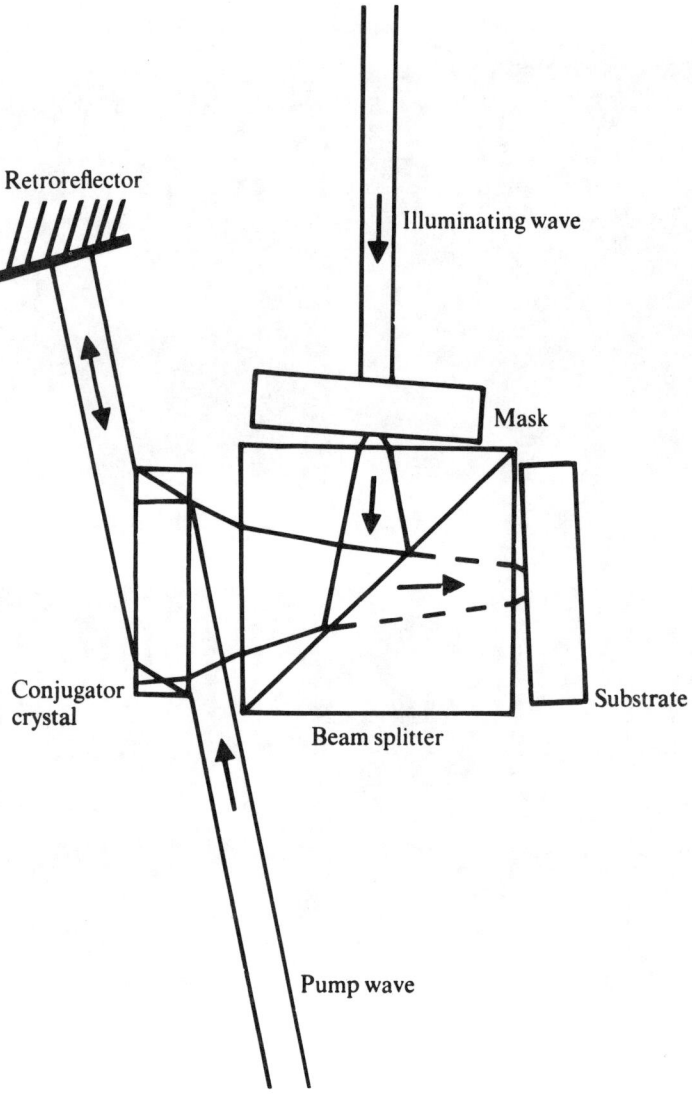

Fig. 3.4 Geometry of the wavefront conjugation lithography apparatus. The pump beams propagate in opposite directions through the nonlinear conjugator medium. The image wave is created upon reflection of the object wave by the medium and its transmission through the beamsplitter. [From Ref. 94]

of producing high-resolution images over large field
sizes. Experiments with various conjugator media have
been reported [94] in which resolution up to 0.5 micron
and exposure times up to 100 s/cm have been obtained.
An example is shown in Fig. 3.5. The practical utility
of this technique is limited by the low efficiency of
available conjugator media and stringent requirements
on the optical surfaces of various components of the
system.

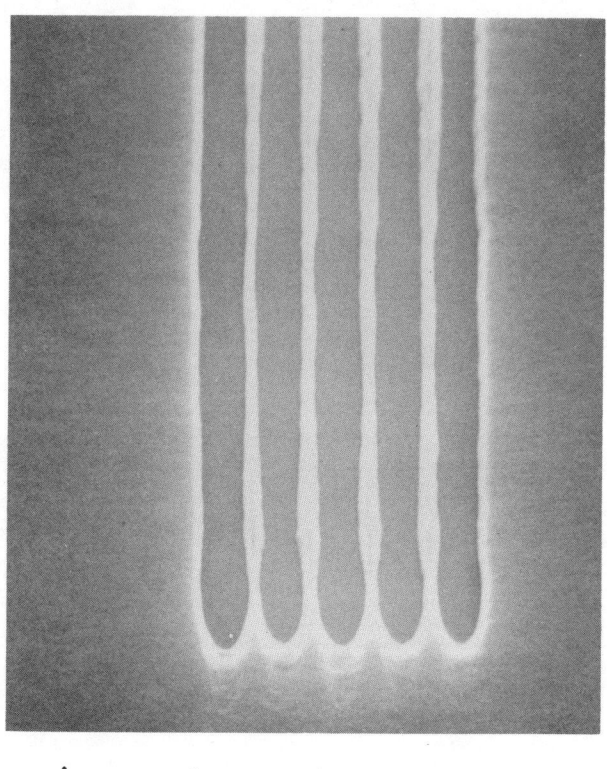

Fig. 3.5   Wavefront conjugation lithography results in
0.9-micron-thick AZ 2400, showing nominally 1-micron-
wide lines and spaces. The photoresist was exposed at
355 nm with the third harmonic of a Nd:YAG laser. The
exposure time was 36 s. [From Ref. 94]

4. Lithography with Excimer Lasers

This chapter describes various concepts for and progress in the implementation of the technology of excimer laser lithography on practical pattern exposure systems. We also discuss here the results obtained in evaluation of different photoresist media for exposure with various excimer laser wavelengths ranging from 308 to 157 nm. Excimer lasers are ideal as light sources for deep ultraviolet (DUV) lithography for a variety of reasons, as discussed previously in Sec. 1.4 and briefly reiterated here. The high output power of these light sources makes it possible to obtain high wafer throughputs and their spatial incoherence helps reduce speckle. The greater power output of excimer lasers also makes more photoresist materials usable in the DUV wavelength region. Optically, excimer lasers may be made to operate with spectral linewidths as narrow as a few picometers, making it possible to design high-performance all-quartz stepper lenses. These attractive features have led to a surge of activity in various aspects of lithography with excimer lasers in the last several years. The results obtained in these efforts have been so promising that it is now widely believed that excimer laser lithography tools will dominate the half- and sub-half-micron wafer fabrication lines in the next decade.

Several lithographic exposure techniques - e.g., contact, full-wafer scanning projection, and step-and-repeat projection - have been explored with excimer lasers. Contact printing is a simple and economical flood exposure technique in which the entire wafer is illuminated in one exposure. However, as a result of repeated contact or near-contact between the mask and the wafer, the mask life becomes severely limited. The technique is most useful for low-volume applications. Full-field scanning is a 1:1 projection technique in which the wafer is exposed through a curved slit long enough to extend across the entire wafer. By scanning the wafer and the mask across the slit, exposure of the full wafer is achieved in one scan. This method

delivers high throughputs and a limiting resolution up
to ~1 micron. In step-and-repeat lithography machines,
the projection lens usually has a reduction ratio of
5:1 or 10:1 and the exposure area is confined to one
chip-field, the wafer then being stepped from chip
to chip. Whereas the throughputs achievable on step-
and-repeat systems are lower than those obtained on
full-wafer scanning tools, the ultimate resolution
capabilities of the former are superior.

In considering application of excimer lasers in
the above exposure systems, the relative importance of
various laser parameters depends on the specifics of
each exposure system. Whereas the high power output of
excimer lasers is advantageous in all exposure methods,
their other characteristics may assume more or less
importance depending on the application. Thus, a high
repetition rate will be a requirement in an excimer
scanning projection tool, but of much less importance
in step-and-repeat projection or contact exposures. On
the other hand, pulse-to-pulse amplitude stability and
narrow spectral bandwidth will be crucial for step-and-
repeat systems, but of much less or no significance in
scanning machines. Along with the differences in the
relative importance of various performance parameters
of the laser, different exposure tools also have widely
varying requirements on the design of their optical
subsystems. These and other criteria will be addressed
below in greater detail when we discuss the results
obtained with various excimer laser lithography tools.

4.1. EXCIMER LASER CONTACT PRINTING

Contact printing is a simple and expedient way to
investigate the imaging characteristics of new light
sources as well as resists. In the first experiments on
lithography with excimer lasers, Jain et al. [1-3] used
contact printing to demonstrate that excellent quality,
speckle-free images can be obtained with resolution
down to 0.5 micron. They used various excimer laser
wavelengths - 308 nm from XeCl, 248 nm from KrF, and
222 nm from KrCl - to delineate patterns in a variety

of resists, including Shipley/AZ 2400, two experimental
resists developed at IBM, and poly(methylmethacrylate)
(PMMA). These images were printed using a brick-pattern
mask with feature (line and space) sizes varying
from 0.5 to 2.0 microns in steps of 0.25 micron. The
chrome mask on a suprasil quartz substrate was held by
vacuum in contact with a 2.5-cm-diameter bare silicon
wafer coated with a 1.0-micron-thick resist film. All
exposures were made with the raw, unprocessed beams
from the excimer lasers. The XeCl laser beam with an
energy of ~50 mJ/pulse was incident on an ~1-cm^2 area
on the wafer. The patterns in AZ 2400 photoresist,
shown in the scanning electron micrographs of Fig.
4.1, were printed using two laser pulses. Thus, the
integrated dose was ~100 mJ/cm^2, delivered in ~20 ns.
After exposure the wafers were developed in a 1:4
solution of Shipley 2401 developer and water. Features
of 1.0 and 0.5 micron dimensions, respectively, are
shown in Figs. 4.1(a) and (b). Note the excellent
image quality and the total absence of speckle. The
latter result, as previously explained in Sec. 2.2.1.3,
is due to the highly random mode structure of the laser
output. This randomness produces an exposure as if it
were carried out by a superposition of an extremely
large number (~10^5) of uncorrelated laser beams.
The speckle contrast, given by the expression $1/\sqrt{N}$,
N being the number of randomly superposed spatially
coherent exposure wavefronts, is therefore negligibly
small. In other words, the above superposition destroys
coherence in a transverse plane and serves to wash out
speckle. To some degree, the large bandwidth (~15 Å),
and hence the poor temporal coherence, of the laser
emission also contributes in the reduction of speckle
in the exposures described above.

We also give an example of images delineated in
AZ 2400 resist using the KrF excimer laser at 248 nm in
the same series of lithography experiments. Figure 4.2
shows 1.5-micron-wide lines and spaces obtained with
an integrated exposure dose of 125 mJ/cm^2, delivered
in five pulses with an attenuated laser output of 25
mJ/pulse. The development was as before for the 308-nm
exposures. Note again the absence of speckle. Further,

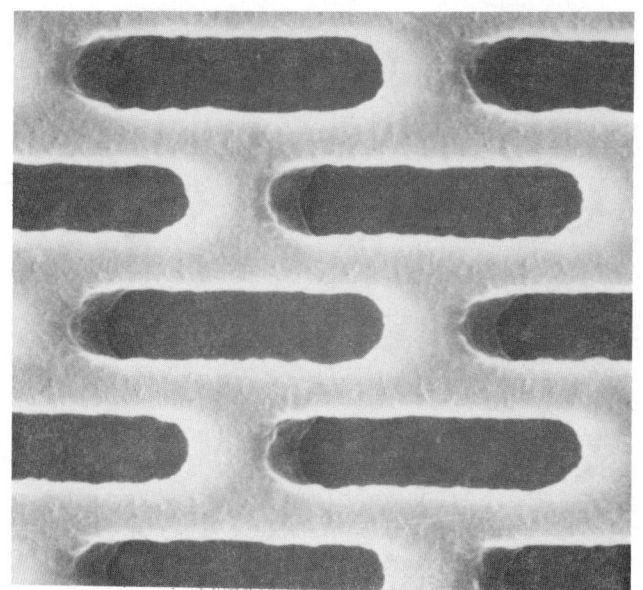

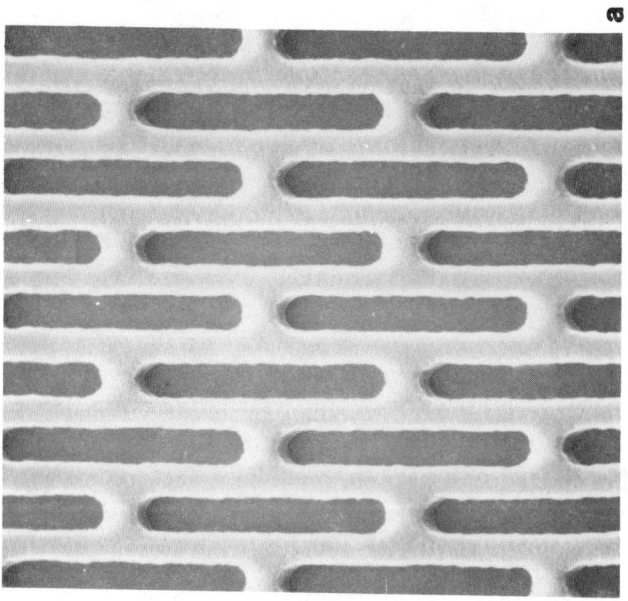

Fig. 4.1 Scanning electron micrographs of images obtained in 1-micron-thick AZ 2400 photoresist with an XeCl excimer laser at 308 nm: (a) 1-micron lines and spaces; (b) 0.5-micron lines and spaces. The exposures were made with two 10-ns wide, 50 mJ/cm² laser pulses. [From Refs. 1-3]

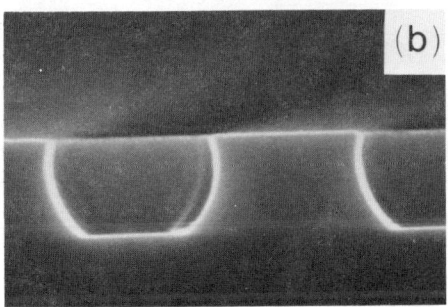

Fig. 4.2   (a) 1.5-micron   lines and spaces obtained in
1-micron-thick   AZ 2400 with a KrF excimer laser at 248
nm. The exposure dose was 125 mJ/cm^2. (b) Cross-section
of (a) showing wall profile. [From Refs. 1-3]

the cross section in Fig. 4.2(b) reveals, as expected, the non-vertical and overcut wall profiles due to the failure of AZ 2400 to bleach adequately when exposed with ∿250-nm deep UV radiation.

Experiments on exposures with the KrF laser in PMMA photoresist have been reported by Kawamura et al. [4]. The energy dose for these large-area pad exposures ranged between 0.085 J/cm^2 and 2.7 J/cm^2. Although PMMA has an extremely poor sensitivity at this wavelength, these authors observed high development rates under certain conditions. This was due, as is discussed in Chapter 6, to the high peak power of excimer laser pulses. Recently, Orvek et al. [27] have investigated the Shipley 4050 and 2400-17 positive-tone resists and the Hitachi 5000P negative-tone resist for exposures with the 248-nm excimer laser wavelength. To give an example, good quality 0.4-micron line/space features obtained by these authors in Shipley 2400-17 resist are shown in Fig. 4.3. Submicron lithography experiments at 248 nm have also been reported by Polasko et al. with an $Ag_2Se/GeSe_2$ inorganic resist [15,19]. In Chapter 6 we describe in detail the varied photoresist response aspects of the above materials.

Several reports have also appeared on excimer laser lithography at wavelengths below 200 nm. Cullmann [22] has used a 193-nm ArF excimer laser exposure of PMMA followed by conventional development in MIBK to print images down to 0.2 micron (Fig. 4.4). Karl-Suss GmbH has announced the commercial availability of an excimer-laser contact/proximity printer. Rice and Jain [12] have shown that direct etching with this excimer wavelength can be used to produce fine patterns in conventional resists; a 0.3-micron line etched in a 1-micron-thick film of Shipley/AZ 2400 resist is shown in Fig. 4.5. In experiments in the vacuum UV (VUV) region with a 157-nm F_2 excimer laser, Craighead et al. [9,23] have produced 0.15-micron line patterns in a trilevel polyimide-Ge-PMMA resist by contact printing using a CaF_2 mask followed by reactive ion etching (Fig. 4.6).

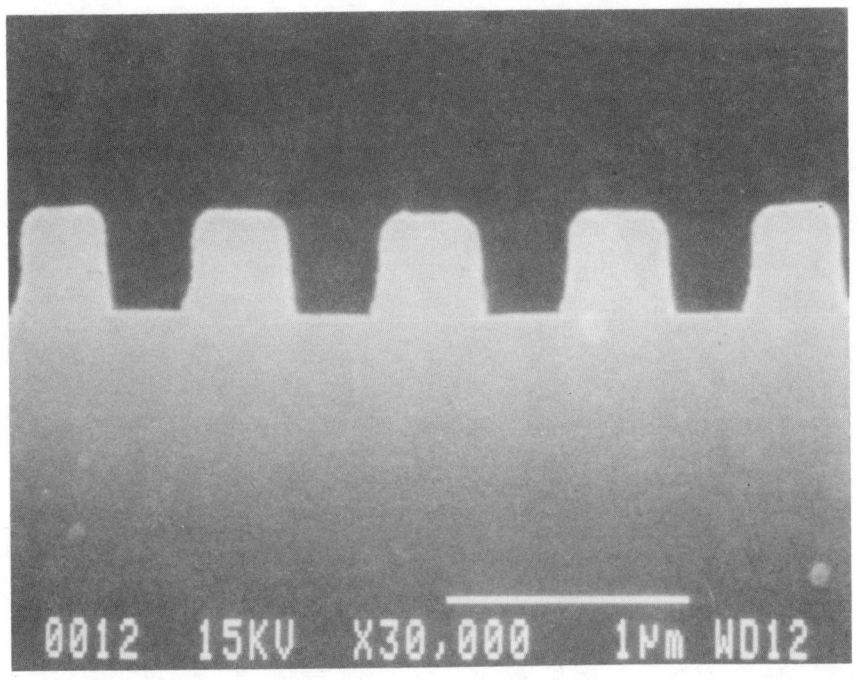

Fig. 4.3   Scanning electron micrograph of 0.8-micron-pitch features in Shipley 2400-17 photoresist contact-printed with a 248-nm KrF excimer laser using an exposure dose of 230 mJ/cm^2. [From Ref. 27]

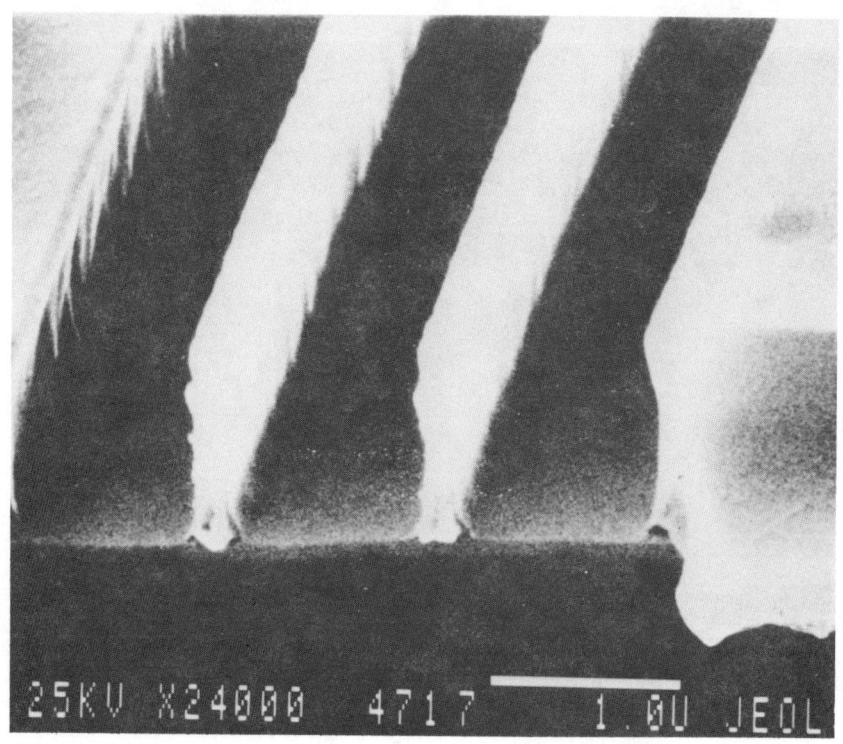

Fig. 4.4   0.2-micron-wide line features contact-
printed in 1.1-micron-thick poly(methlmethacrylate)
(PMMA) with 193-nm ArF excimer laser exposure.
[From Ref. 22]

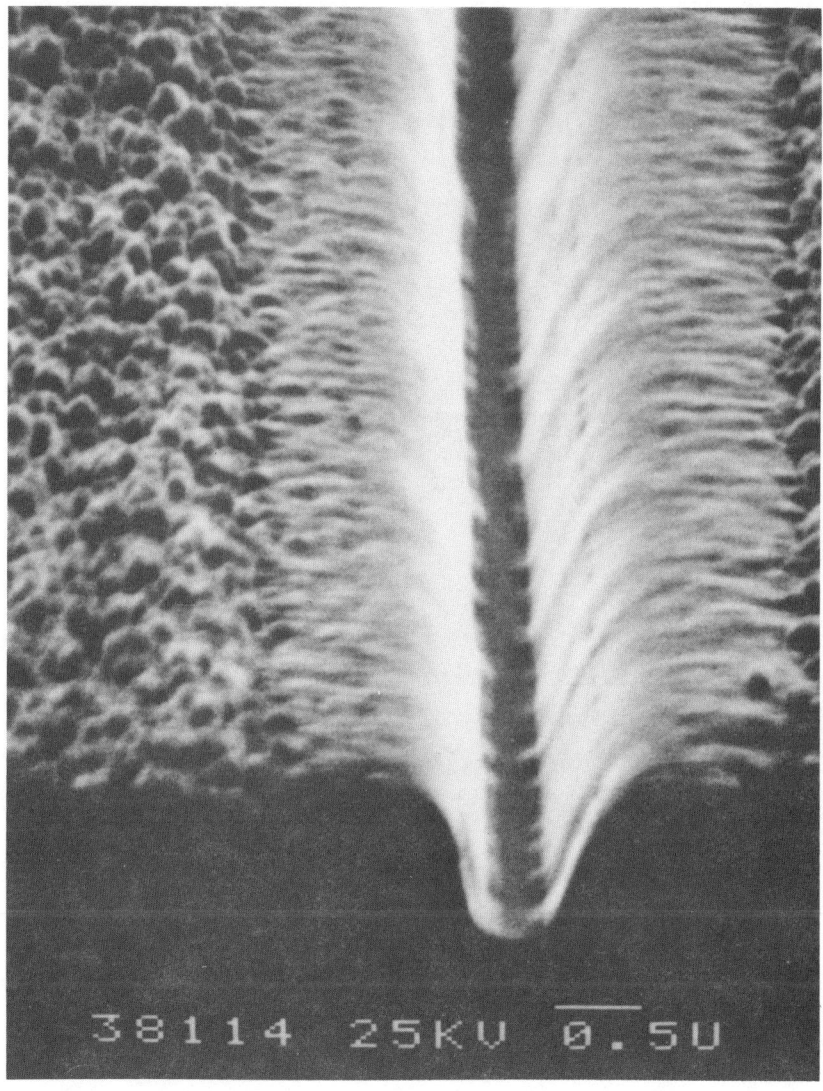

Fig. 4.5 A 0.3-micron line directly etched in AZ 2400 resist by contact exposure with a 193-nm ArF excimer laser. The total dose was 1 J/cm^2, delivered with pulses of 13 mJ/cm^2 energy. [From Ref. 12]

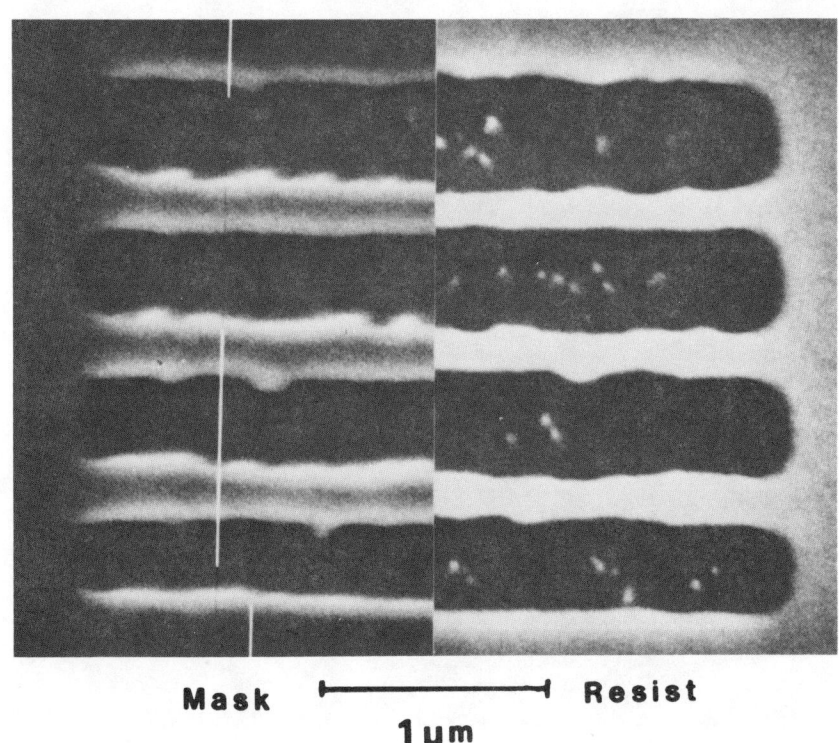

Mask ├───────────────┤ **Resist**

1 µm

Fig. 4.6   Images obtained in a   trilevel polyimide-
Ge-PMMA resist by contact printing with a vacuum UV
F_2 excimer laser followed by reactive ion etching.
[From Refs. 9,23]

4.2. EXCIMER LASER PROJECTION LITHOGRAPHY

To be commercially attractive, any new optical imaging technology must be demonstrated on a projection system; therefore, several successful efforts have been made to develop a variety of excimer laser projection lithography systems to print micron- and sub-micron-size images. These include a number of commercial tools as well as laboratory setups of different types. In the following discussion, we review the development of full-field scanning and step-and-repeat excimer laser projection systems and describe the results obtained on both. In addition, we also describe an excimer laser pattern generator for mask making.

4.2.1. Full-Field Scanning Projection Lithography with Excimer Lasers

Excimer laser projection lithography on full-field scanning tools has been demonstrated on Perkin-Elmer Model Micralign 111 and Model Micralign 500 machines by Jain et al. [8,11,29]. The Perkin-Elmer scanners are ring-field, f/3 optical systems in which a 1:1 image of an annular segment of the mask is projected by a set of mirrors on the wafer. Fig. 4.7 shows a schematic of the projection system of the Model 500, in which the annular segment has a width of 4 mm, a radius of 115 mm, and a length of 135 mm - sufficient to accommodate 125-mm-diameter wafers. Exposures on the full wafer are made by simultaneously scanning the mask and the wafer across the annular segment. Due to the cancellation of the third- and fifth-order optical aberrations [95], 1-micron resolution at exposure wavelengths in the vicinity of 300 nm is readily achieved over the entire length of the segment. In the conventional mode of operation with a mercury lamp, the arc-shaped illumination on the mask is produced by the condenser unit of the Perkin-Elmer system. The condenser is an anamorphic optical system that uses a curved mercury arc lamp and appropriate optical components, both reflective and refractive (see Fig. 4.8), to produce the illumination characteristics on the mask with the required shape and numerical aperture. The arc source

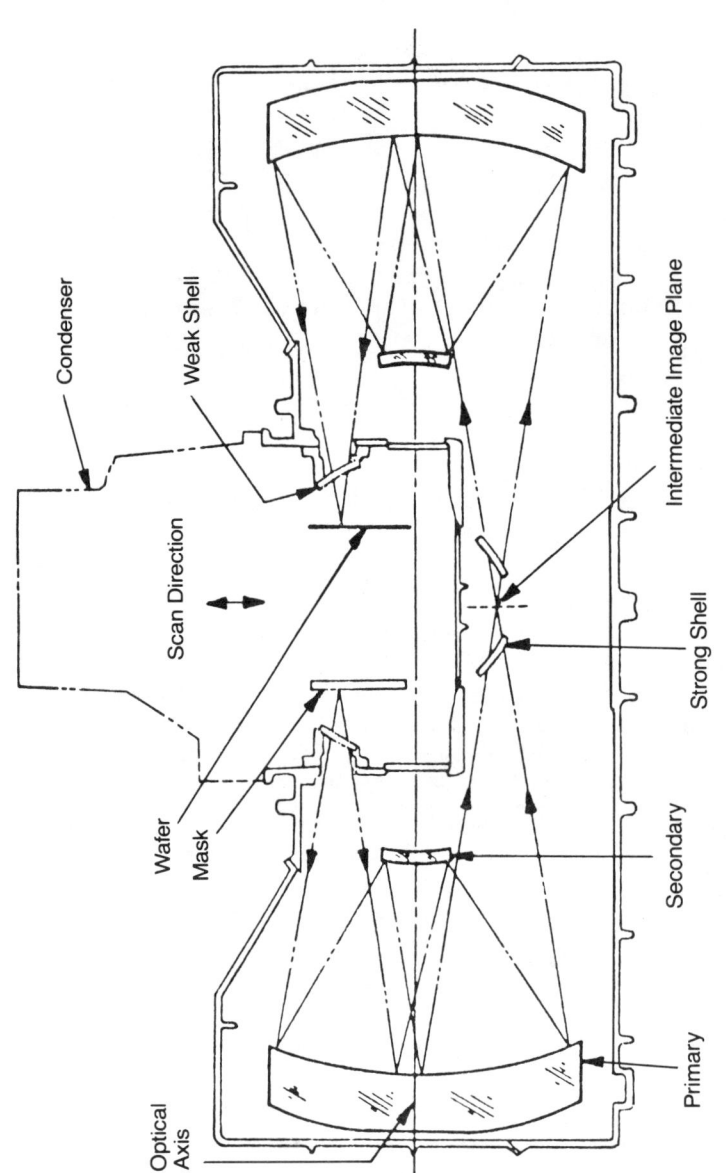

Fig. 4.7   Schematic of the projection optics in the Perkin-Elmer Micralign 500 lithography system. Note the 1:1 imaging and the simultaneous scanning of the mask and the wafer. [From Ref. 29]

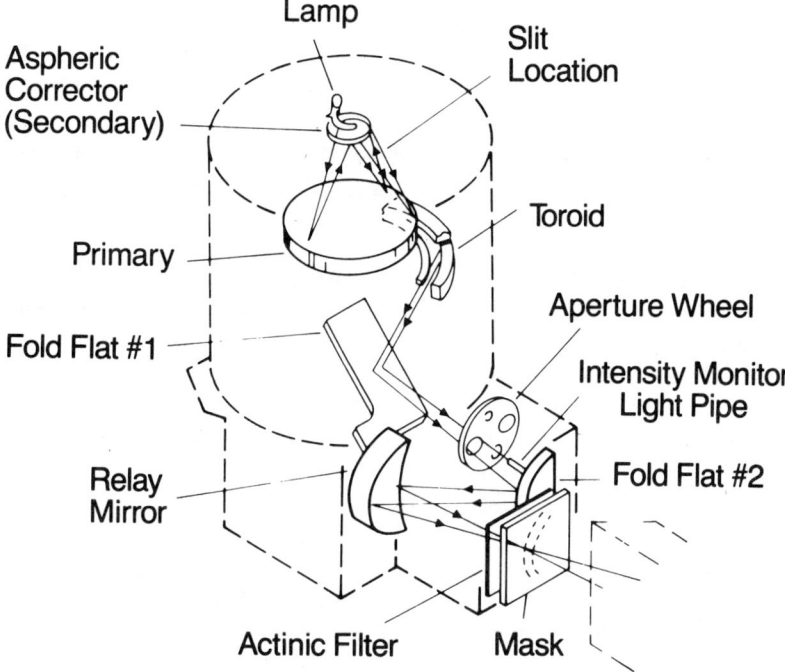

Fig. 4.8   Schematic of the condenser optics employed in the Perkin-Elmer Micralign 1:1 full-wafer scanning lithography systems. Note the curved capillary lamp and the aspheric imaging components. [From Ref. 29]

thus produced in the mask plane matches the zone of optimum correction for the projection system.

For excimer laser exposures on a projection system of the above type [29], the laser beam must be suitably transformed so as to simulate, as seen by the mask, the illumination characteristics of the Perkin-Elmer lamp-condenser unit. The key features of the optical system of such a condenser are: (a) it generates a curved arc at the mask plane; (b) the arc is effectively self-luminous; (c) each point on the arc radiates into a uniform and well-defined numerical aperture; and (d) the illumination is telecentric and uniform. The task of the transformation system is then to reconfigure the laser radiation so that it satisfies the above four criteria.

There are several techniques for transforming a collimated, rectangular laser beam into a source with the above characteristics. The simplest of these is a method that uses anamorphic optics and a diffuser [29, 96]. It is illustrated schematically in Fig. 4.9. The combination of lenses L1 and L2 is employed as a beam expander or contractor, and mirror M1 as a steering mirror. The cylindrical lens L3 focuses the rectangular beam into a fine line at S'. A cylindrical mirror M2 is inserted at oblique incidence in the beam path before S' and, as a result, arc S is produced at the lamp plane. The lamp is removed and replaced by a quartz diffuser D placed in the lamp plane, thus transforming the laser radiation into an effectively self-luminous arc-shaped source. The aperture selection wheel in the condenser unit is used to select the appropriate NA of the illumination at the mask. It may be remarked here that, with the projection NA being fixed, selection of the condenser NA permits one to select the partial coherence σ of the system, σ being given by the ratio of the condenser NA to the projection NA.

Excimer laser projection lithography experiments were carried out on a system transformed in the above manner using an XeCl laser operating at 308 nm at a repetition rate of 100 Hz and producing an output power

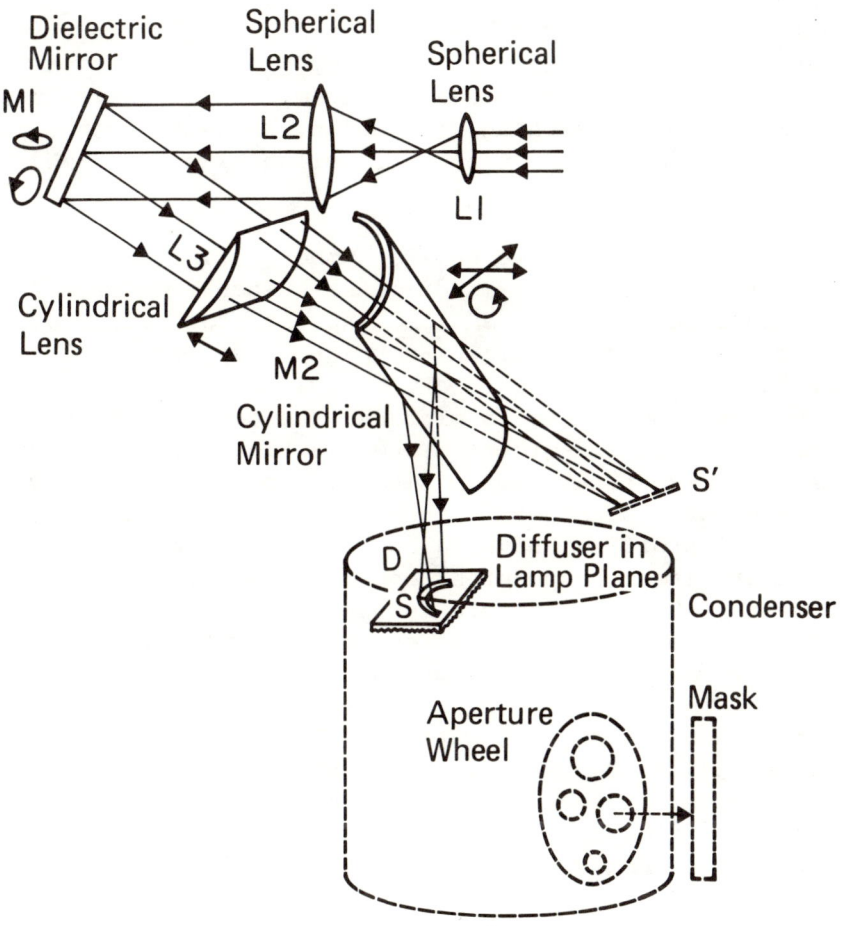

Fig. 4.9   Anamorphic optical system for transforming the collimated rectangular beam from an excimer laser into a self-luminous arc with a specified numerical aperture. [From Ref. 29,96]

in the vicinity of 10 W. The numerical aperture in the
condenser was set at 0.087, giving a partial coherence
factor σ = 0.087/0.167 = 0.52 for the exposures. The
optical efficiency of the laser beam transformation
assembly was estimated to be $\sim$20-25%. Since the
efficiency of the Perkin-Elmer condenser and projection
systems is each in the vicinity of 25%, the optical
throughput of the overall system, i.e., the fraction
of laser power reaching the wafer, was $\sim$6%. Thus, the
power incident in the wafer plane was $\sim$600 mW, or
$\sim$100 mW/cm^2.

In Fig. 4.10 we show a number of images obtained
in a full-wafer (125-mm diameter) scan time of 27 s.
The photoresist, ER1, an IBM experimental formulation
of the diazonaphthoquinone-novolak type and similar to
AZ 2400, was 1 micron thick. The various features range
in size from 1 to 5 microns. The cross sections of
the 1.0- and 1.5-micron lines in Fig. 4.10(c) depict
the near-vertical (85°) sidewall profiles that were
consistently obtained in these exposures. Similar near-
vertical profiles have also been obtained in 308-nm
excimer laser exposures on a Perkin-Elmer Model 111
system [29]. An example is shown in Fig. 4.11. Note
that conventional lamp exposures with the 313-nm
mercury line typically produce images with sidewall
angles in the 50-70° range due to the normal optical
modulation transfer function (MTF) considerations as
discussed in Sec. 1.3. Thus, it may be claimed that
the above excimer laser exposures produce images
with significantly improved MTF. A number of possible
explanations have been considered for this improvement.

The high-pressure mercury lamp employed in the
conventional illumination has several spectral peaks in
the mid-UV and DUV regions. In addition to the desired
313-nm line, a significant fraction of the emission at
two adjacent peaks, 297 and 254 nm, reaches the wafer
due to the broadband optics used in the Perkin-Elmer
systems. It is conceivable that these short wavelengths
are primarily absorbed in the upper portions of the
photoresist film causing overexposure in that region.
This may result in a faster development rate near the

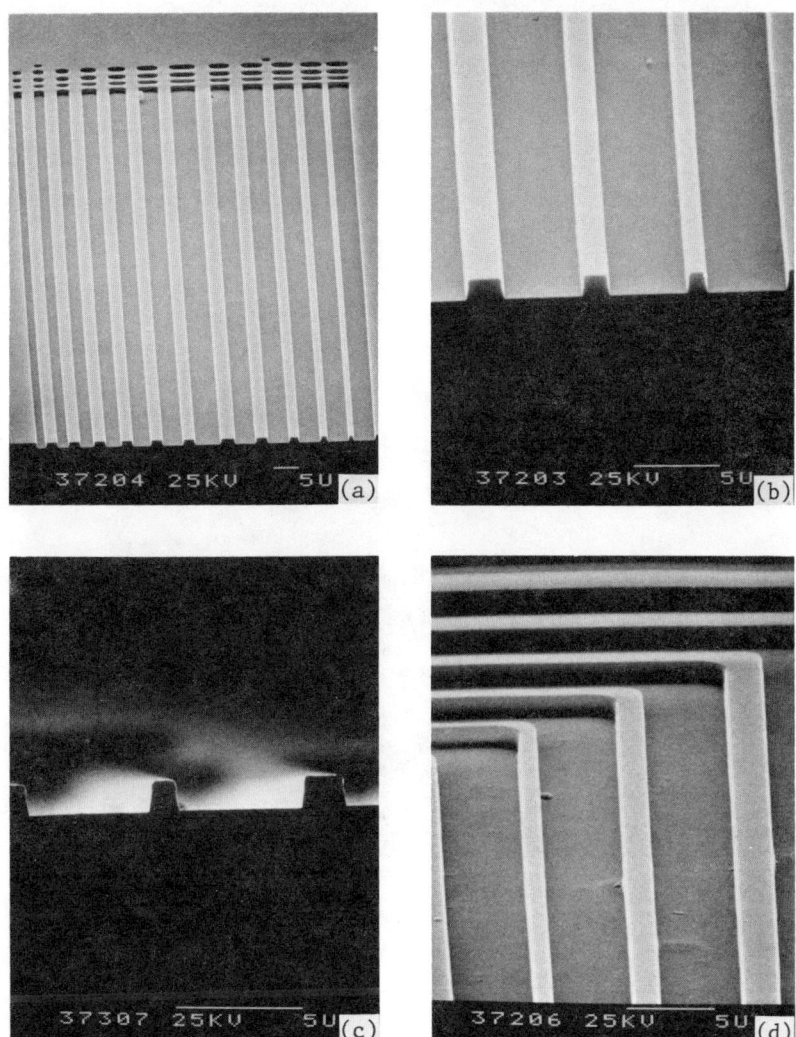

Fig. 4.10   Projection lithography on the Perkin-Elmer
Micralign 500 system with a 308-nm XeCl excimer laser:
(a) lines of width from 1 to 3 microns and spaces of
width from 1.5 to 5 microns. (b) Enlarged view of the
1.0-, 1.5-, and 2.0-micron lines. (c) Cross sections of
the 1.0- and 1.5-micron lines showing near-vertical
(85°) wall profiles. (d) Images showing the resolution
fidelity at corners of the 1.0-, 1.5-, and 2.0-micron
lines. [From Ref. 29]

Fig. 4.11   Images printed on a Perkin-Elmer Micralign
111 projection lithography system with a 308-nm XeCl
excimer laser, showing lines of width from 1 to 3
microns. An optical coupling system similar to that
shown in Fig. 4.9 was used. [From Ref. 29]

top of the resist and therefore more gradually sloped walls. To examine whether the monochromaticity of the laser illumination produces the steep profiles of Fig. 4.9 by eliminating the shorter wavelengths, exposures were made [29] on the Perkin-Elmer Model 111 system with the following bandwidth filtering: (a) broadband illumination with the unfiltered lamp; (b) illumination filtering peaked at 311 nm with half maxima at 287 and 335 nm; and (c) illumination filtering at 314 nm with half maxima at 308 and 320 nm. The reduction in the wafer-plane energy density with increased filtering was compensated by using proportionately longer exposure times. In all cases the development conditions were kept constant. In each of these exposures, identical 53° wall profiles were obtained, indicating that in the wavelength region explored above the influence of the illumination bandwidth is at best minor.

Since the instantaneous power densities used in the excimer laser exposures are very different from those in conventional lamp exposures, one may expect differences in the resist behavior due to possible nonlinear photochemical reactions at high powers. The peak power densities in excimer laser exposures are in the neighborhood of 5 MW/cm^2, whereas in the lamp case, the peak power density is a fraction of 1 W/cm^2, i.e., $\sim 10^7$ times less. To investigate the influence of this difference systematically, exposures were made using the excimer laser with several reduced peak powers by successively attenuating the laser beam by factors of ten [29]. For each lower peak power, correspondingly longer exposure times were used to produce images with identical exposure energy doses. Images were produced with the peak power density ranging from 100% of the maximum available (~ 5 MW/cm^2) to 0.01% of the maximum (~ 500 W/cm^2). In each case the image wall profile was measured from cross-section views in scanning electron micrographs. Further lowering of the peak power would have necessitated exposures many hours long, during which time the stability of the projection tool would be questionable. In the power range investigated in the above work, no significant difference was found in the image profiles, i.e., they were in the 80-85° range

in all cases. Thus, although due to limitation of the
exposure tool used it was not possible to provide clear
evidence of resist-response nonlinearities, it may be
concluded that the onset of any such processes should
occur in the 10^0-10^2 W/cm^2 region. It was also found
that even though the image profiles indicate certain
nonlinear exposure mechanism in the photoresist, the
dependence of the exposure dose on the incident peak
power showed no nonlinearity. These and several other
aspects of photoresist behavior under excimer laser
illumination are investigated at greater length in
Chapter 6.

The excimer laser implementation of full-wafer
scanning tools as described above has also been used to
demonstrate deep UV excimer laser projection printing.
Images produced on a Perkin-Elmer Model 111 machine
using a 248-nm KrF laser and the optical transformation
system of Fig. 4.9 are shown in Fig. 4.12. These images
were made in a 1-micron-thick film of Shipley/AZ 2400
resist. The sloped wall profiles in these exposures are
a result of the excessive absorption of AZ 2400 in the
deep UV region, as mentioned previously in Sec. 4.1.

The annular-ring-field imaging system, which forms
the basis of the full-wafer scanning projection tools,
has also been used in a simplified, laboratory version
as an excimer laser projection lithography apparatus by
Nakase [20]. This arrangement, along with examples of
images printed with the 308-nm XeCl laser in OFPR 800
resist, are shown in Fig. 4.13. Finally, it should be
remarked that in addition to Perkin-Elmer, Canon also
manufactures 1:1 ring-field full-wafer scanning tools,
and the entire discussion in the above paragraphs on
excimer laser projection lithography is applicable to
them as well.

We now discuss the relative importance of various
excimer laser parameters as they relate to lithography
on scanning systems. A laser power output of 5-10 W
is adequate to give wafer exposure throughputs in the
vicinity of 100 125-mm-diameter wafers/h. The laser
spectral linewidth is not critical, i.e., the laser

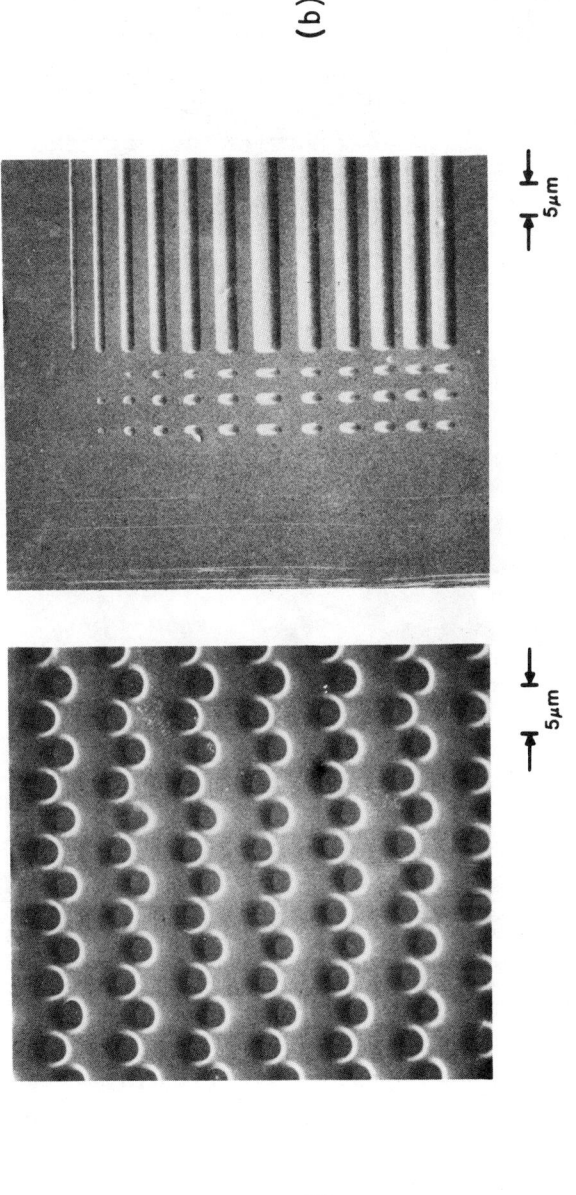

(b)

(a)

5μm

5μm

Fig. 4.12   Deep UV projection lithography with a 248-nm KrF excimer laser on a Perkin-Elmer 1:1 full-wafer scanning system. An 82-mm-diameter wafer coated with 1-micron-thick AZ 2400 resist was exposed in 30 s. The optical coupling system was similar to that shown in Fig. 4.9. (a) 3-micron-diameter holes separated by 1-micron walls. (b) Lines of width from 1 to 5 microns and spaces of width from 1.5 to 5 microns. [K. Jain and R. T. Kerth, unpublished results]

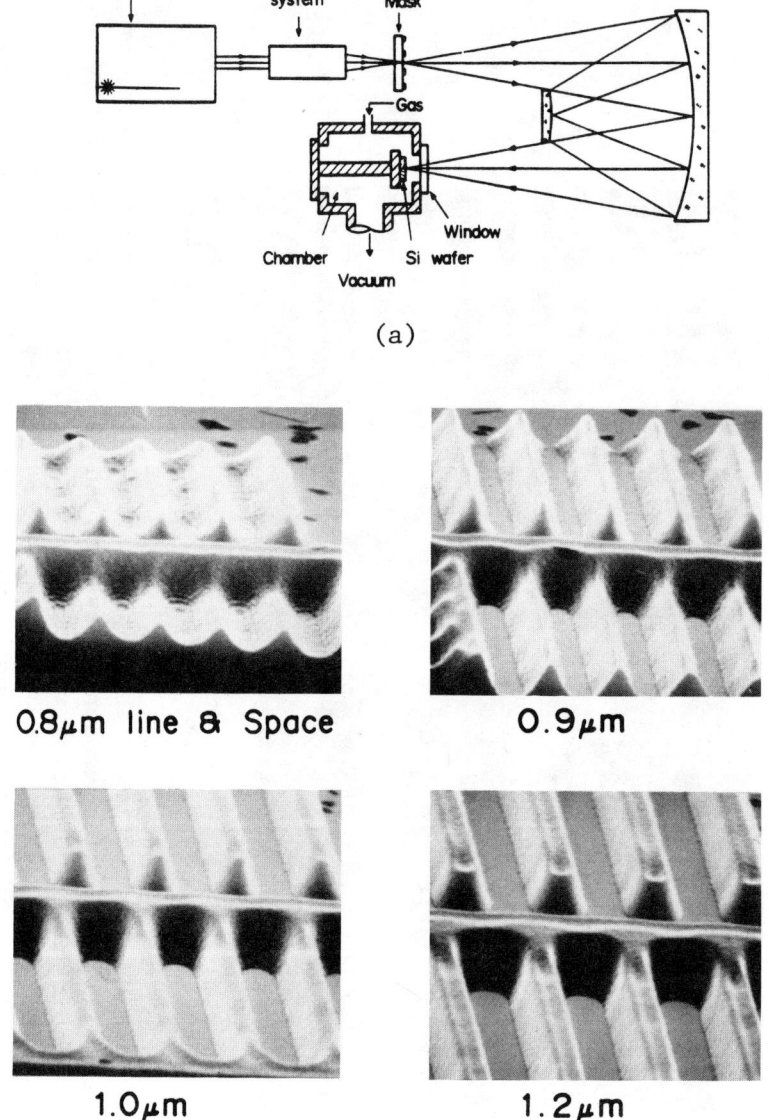

(a)

0.8μm line & Space   0.9μm

1.0μm   1.2μm

(b)

Fig. 4.13   (a) Schematic of a 1:1 exposure system
using an excimer laser source. (b) Images   exposed in
1.4-micron-thick OFPR resist with a 308-nm XeCl laser
in the system of (a). [From Ref. 20]

need not be narrowed spectrally, because the bandwidth
of the ring-field systems, which are mostly reflective,
is far larger than the raw excimer linewidth, which is
typically on the order of 5-10 Å. On the other hand,
a high laser pulse repetition rate is desirable, so
that exposures may be produced with overlap of several
pulses. As an example, in the exposures on the Perkin-
Elmer Model 500 described above, the excimer laser had
a repetition rate of 100 Hz, so that with a slit width
of 4 mm and a scan time of 27 s, each point on the
wafer was exposed with 85 overlapping pulses. Such an
overlap eliminates the need not only for high pulse-
to-pulse amplitude stability but also for uniformity
of the pulse envelope in the scan direction.

     To produce the ring-field illumination required in
scanning systems of the type described above, methods
other than the anamorphic-optics-diffuser system of
Fig. 4.9 have also been developed. Before discussing
them, it should be emphasized that these optical beam
transformation systems have a much wider applicability
than only for microlithography on Perkin-Elmer-type
full-wafer 1:1 scanners. Thus, any optical illumination
system in which a field shape other than rectangular
must be created with a certain NA will require a beam
transformation system of some sort.

     A fiber optic technique [97,98] for illumination
transformation is shown in Fig. 4.14. The laser beam,
after being suitably expanded by a beam expander, fills
a rectangular, two-dimensional fly's-eye lens array,
which divides the input beam into a large number of
smaller beams. Each of these small beams is focused by
a lenslet, i.e., an element of the lens array, and made
to enter one of the fibers in the input face of a fiber
bundle. The fiber faces at the input of the bundle are
arranged so as to map the lens-array elements one-to-
one. The focal length of the lens elements is chosen
to produce the required numerical aperture of each of
the multitude of beamlets. Since transmission of the
laser radiation within the fibers is by total internal
reflection, this numerical aperture is maintained at
the exit face of the fiber bundle. The fibers at the

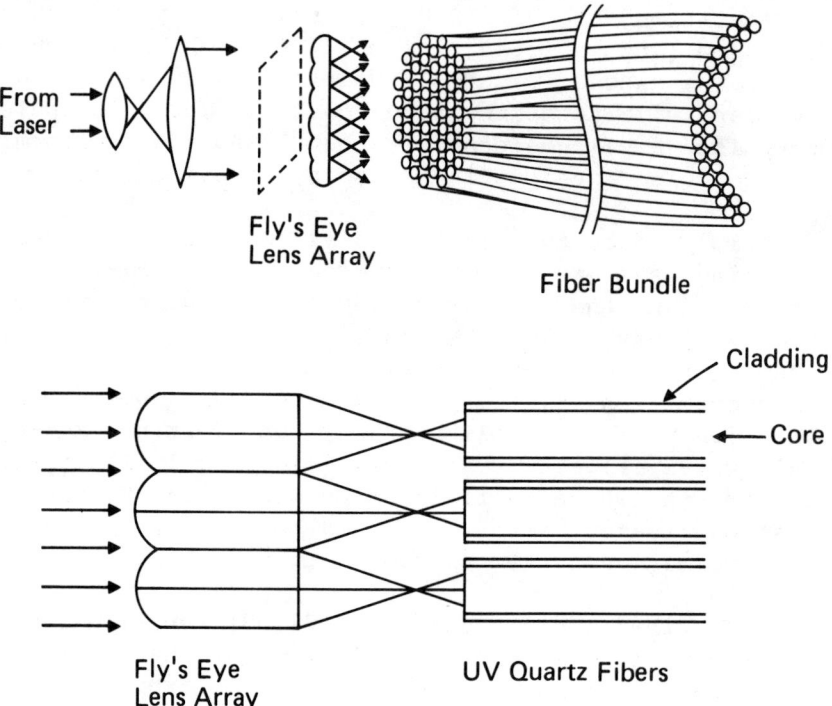

Fig. 4.14   A fiber optic coupling system for transforming an excimer laser beam into a self-luminous arc
with a specified numerical aperture [From Refs. 97,98]

exit are arranged in an arc to produce the required width and curvature of illumination at the mask plane. At an optimum distance from the exit face, the cones exiting from different fibers overlap partially so that a uniform, self-luminous, arc-shaped illumination with the required numerical aperture is produced. One may readily see that, with appropriate fiber distributions, output shapes other than an annulus can be produced. Further design details and variations can be found in Refs. 97 and 98.

Figure 4.15 shows a holographic beam transformation system to produce ring-field illumination [99]. This interesting concept, on the one hand, is more elegant than the anamorphic optics and the fiber optics systems described above, but on the other hand, it is also more difficult to implement. The basic underlying principle is to record a hologram, which, when played back with a collimated beam, would reconstruct the desired arcuate wavefront. Since the conventional excimer laser beam is not sufficiently coherent for recording holograms, one must use either an injection-locked excimer laser or a frequency-upconverted dye or Nd:YAG laser emitting at the wavelength of the excimer laser. This coherent collimated laser beam is divided into two beams with a beamsplitter. One of the two beams impinges on the recording medium directly. The other beam illuminates a diffuser plate on which only an arc-shaped area has been left unmasked. Thus, the two sources made to interfere in the recording medium to produce the hologram are a collimated beam and a self-luminous arc. In playback, when such a hologram is illuminated by the conjugate of the collimated beam (which now may come from a conventional excimer laser), an arcuate source will be produced with the required radiation characteristics. The hologram may be either reflective or transmissive. Reference 99 describes various details of the method. Again, as in the fiber optics case, the holographic system is applicable generally, and any desired illumination characteristics may be produced.

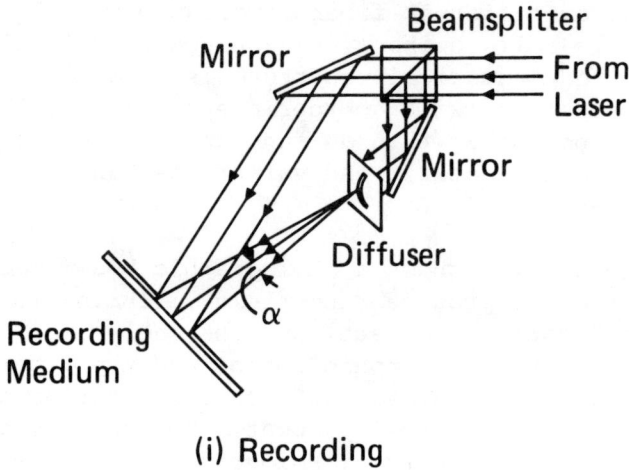

(i) Recording

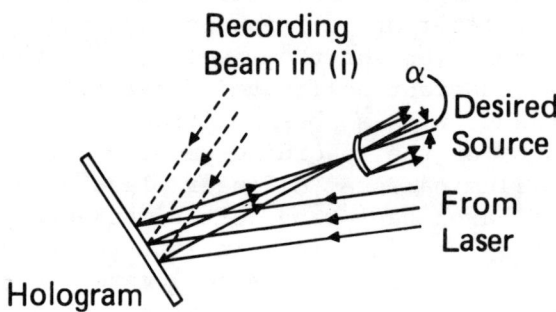

(ii) Reconstruction

Fig. 4.15   A holographic beam-transformation technique
to produce ring-field illumination with a specified
numerical aperture [From Ref. 99]

4.2.2. Step-and-Repeat Excimer Laser Projection Lithography

Step-and-repeat projection printing with various reduction ratios (typically 1, 5, or 10) has evolved during the 1980s as the dominant force in optical lithography in the sub-micron region. With the high power output and small spectral linewidths that excimer laser sources are able to provide at deep and vacuum ultraviolet wavelengths, it is likely that excimer laser step-and-repeat lithography systems will emerge as the primary technology for production lithography in the half- and sub-half-micron regimes.

Step-and-repeat imaging differs conceptually from 1:1 full-field scanning projection and, therefore, the required laser source characteristics and other system requirements are also different in the two exposure techniques. In refractive systems, which constitute the majority of steppers, proper image correction over the desired exposure field requires use of a large number of lens elements as well as use of different optical materials for lens fabrication. At DUV wavelengths, all glasses become unusable as possible lens materials due to their poor transmission, and one is left with only a limited number of acceptable candidates, which include quartz, various fluoride crystals, and sapphire. The fluoride crystals - lithium fluoride (LiF), magnesium fluoride (MgF_2), barium fluoride (BaF_2), and calcium fluoride (CaF_2) - all have good optical transmission in the DUV, but possess poor mechanical rigidity, making them unsuitable for fabrication of optical elements with high-quality surface finish requirements. Further, CaF_2 is also undesirable because of its poor thermal conductivity, and MgF_2 because of its birefringence. Sapphire has good mechanical and thermal properties, but its transmission in the DUV is not sufficiently high. Thus, quartz is left as the only suitable material for fabrication of high-resolution projection lenses in the 250-nm region.

This drastically reduced choice of materials, in turn, places a stringent requirement on the maximum usable laser linewidth; for most stepper lens designs, the laser line can be no wider than 0.004 nm. A narrow linewidth also implies that the center frequency of the laser must be stabilized to within a fraction of the linewidth. Further, since a stepper typically exposes each field with fewer pulses than the number of pulses that produce an overlapping exposure in a full-field scanner, pulse-to-pulse amplitude stability as well as spatial uniformity of the laser beam are of greater importance in the case of steppers. On the other hand, a high pulse repetition rate of the laser, although always desirable, is a more critical parameter for scanning applications. Finally, it is useful to note that, although spectral narrowing of the laser reduces its temporal and spatial incoherence, reappearance of speckle in the images is generally not a concern in excimer laser steppers due to the averaging-over-pulses that results from uniformization of the beam as well as from multiple-pulse exposure.

Before proceeding to discussions of various actual step-and-repeat systems designed with excimer laser illumination, it is interesting to note that with the availability of excimer laser sources, it is now not unrealistic to envision a flash-on-the-fly projection exposure system. Essentially, such a system would be optically a step-and-repeat concept, but in which the stage, rather than stepping from one field to the next, would move continuously. Such an exposure technique is realistic only with excimer laser sources because only they can provide a light pulse that has a sufficiently narrow pulsewidth so that the stage movement during its duration is negligible and sufficiently intense so that it can expose an entire field. Note that pulse-to-pulse uniformity of the laser will be critically important in a practical realization of a flash-on-the-fly machine.

A laboratory arrangement of the step-and-repeat type for carrying out deep UV projection lithography experiments with the KrF laser is shown in Fig. 4.16 [7]. This small-field (~ 1 mm^2) system used a 0.2 NA microscope objective as the imaging lens. Although, as

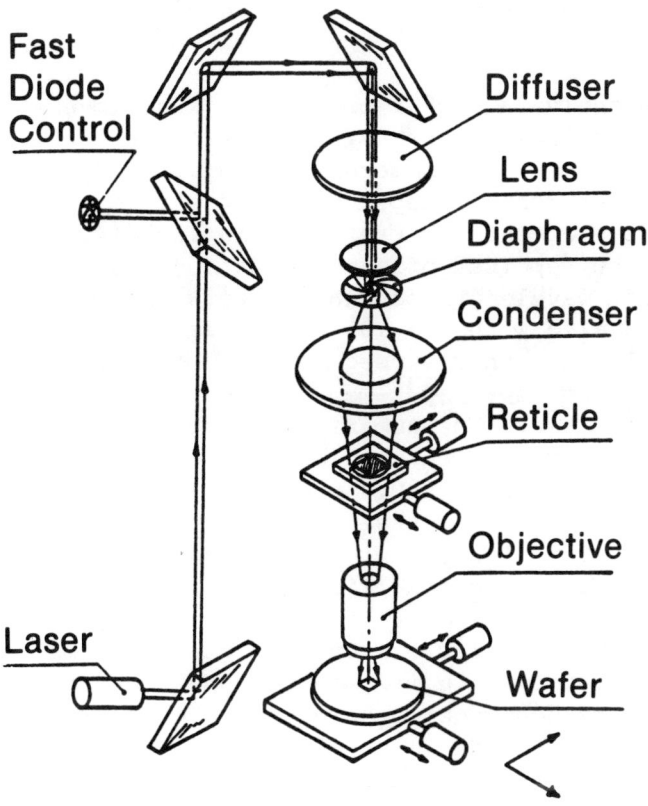

Fig. 4.16   Schematic of an experimental 10:1 reduction projection system for use with a KrF excimer laser at 248 nm.   The projection lens consists of a microscope objective with an NA of 0.2. [From Ref. 7]

might have been expected, the performance of the
apparatus was severely limited due to aberrations and
distortions in the lens, reasonable imaging in the 1-
micron regime was demonstrated. The nonbleachability of
the resist used, Shipley/AZ 2400, was also responsible
for degradation of the image profiles. Some examples of
the exposures reported in these experiments are shown
in Fig. 4.17.

Recently, a number of efforts have been reported
on the implementation of excimer laser illumination
systems on various commercially built steppers as well
as on experimental step-and-repeat systems. A survey of
these developments is presented in Table 4.1. Although
all of the reported efforts to date have been at the
KrF laser wavelength of 248 nm, work is known to be in
progress also with the 193-nm ArF laser. Most systems
in Table 4.1 have used a reduction ratio of 5:1;
however, some 10:1 systems and one 1:1 system have also
been reported. The numerical aperture of the various
systems has ranged between 0.3 and 0.42. We remark here
that an NA value of 0.4 produces a resolution, w, of
0.5 micron if one uses in the equation $w = k \lambda /NA$ a
value of 0.8 for the process parameter k. The 5:1 and
10:1 reduction lenses listed in Table 4.1 have been
designed with exposure field sizes varying from less
than 10-mm diameter in experimental systems to ~15x15
mm^2 in prototype units. Although some of the lens
designs are nonmonochromatic and have incorporated both
quartz and fluoride lens elements to permit the use of
a non-line-narrowed excimer laser, most are monochromatic, with only quartz lens elements and requiring
a spectrally narrowed laser. The line narrowing of the
excimer laser has been accomplished by the use of
various frequency-selective elements, such as etalons,
gratings, prisms, and other dispersive elements.
Finally, the alignment systems used in these efforts
have not been specifically designed for DUV excimer
laser steppers; rather, they are adaptations of the
alignment systems found in various existing, conventional, non-excimer commercial machines. In the next
few paragraphs, some of the developments summarized in
Table 4.1 are discussed in greater detail.

Fig. 4.17 Images obtained in AZ 2400 resist with the deep UV excimer laser projection system of Fig. 4.16. Note the sloped image profiles caused by the strong and nonbleaching absorbance of AZ 2400 in the deep UV. [From Ref. 7]

Table 4.1   Survey of reported developments in excimer laser steppers.

	Wavelength (nm)	NA	Reduction Ratio	Field Size (mm)	Resolution (μm)	Align. Tol. (±μm)	Lens Type	Lens Optics (nm)	Laser Δλ (nm)	Line Narrowing Technique
AT&T	248	.20-.38	5	14.5-20φ	.52	.25(2σ)	mono	Quartz	.005	Etalon
GCA	248	.35	5	20φ	.57	.2(3σ)	mono	Quartz	.004	Dispers.
Matsushita	248	.37	5	15x15	.54	.3(3σ)	mono	Quartz	.005	Etalon
Toshiba	248	.37	10	5x5	.54	.2(3σ)	achro	Q+CaF$_2$	.4	–
Nikon	248	.42	5	15x15	.47	.18(3σ)	mono	Quartz	.003	?
Canon	248	.35	5	5φ	.57	?	mono	Quartz	.4	?
Mitsubishi	248	.37	10	5x5	.54	?	achro	Q+?	.4	–
Sony	248	.35	5	4x4	.57	?	mono	Quartz	.002	?
Mitsui	248	.30-.40	?	21.2φ	.50	?	mono	Quartz	.007	Etalon
NTT	248	.42	5	10x10	.47	?	mono	?	.007	?
Rutherford	248	.30	1	40x15	.66	?	achro	Q+LiF	.3	–

A schematic illustration of an excimer laser step-and-repeat lithography apparatus developed at AT&T is shown in Fig. 4.18. This prototype unit consisted of a commercial GCA Model 4800 DSW wafer stepper retrofitted with a 248-nm KrF laser illumination system and a DUV projection lens [26]. The lens had a reduction ratio of 5:1, a numerical aperture of 0.38, and an exposure field diameter of 14.5 mm. The lens was made of fused silica elements only and, thus, had no chromatic correction. The required line-narrowing of the laser output, to 0.005-0.007 nm, was achieved by use of two intracavity etalons. The resist used was Shipley Microposit 2400-17 at a thickness of 500 nm. Due to excessive absorption in the DUV, this resist was used as the imaging layer in a conventional trilevel structure. Fig. 4.19 shows 0.5-micron-wide patterns etch-transferred into 1.8-micron-thick Hunt HPR 206 resist. In the above system the line-narrowed laser output was ~5% of the power available without narrowing. Such a poor conversion efficiency is typical of the above method of spectral narrowing; significant improvement can be made by using an injection-locked laser system. Additional losses in the optical system - primarily due to inefficient diffusing elements used as uniformizers - reduced the average power reaching the wafer to 4 mW/cm^2, thereby necessitating a long exposure time of 25 s per site. An optimized optical system used with an injection-locked laser would be able to reduce the site exposure time significantly. A step taken in this direction has been reported recently [36] in which, with the use of an injection-locked system, the power in the wafer plane was increased to ~40 mW/cm^2, reducing the site exposure time to ~3 s.

The GCA Corporation has developed two different monochromatic excimer laser stepper tool prototypes by modifying its existing Model DSW 8000 system offered at the conventional Hg g-line (436 nm) and i-line (365 nm) wavelengths. Both use the 248-nm KrF laser and all-quartz reduction lenses [42,51]. One of them has a 10:1 lens with a numerical aperture of 0.35 and an exposure field diameter of 14 mm, the other being a 5:1 system with a 0.35 NA and a 20-mm field diameter. The laser is

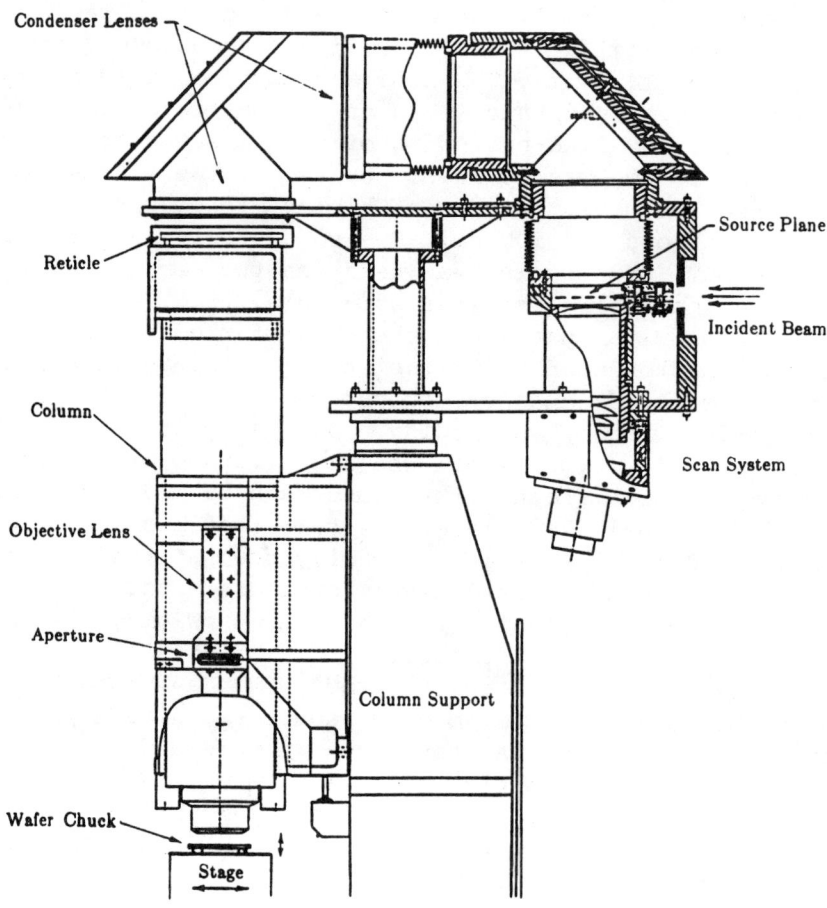

Fig. 4.18   Schematic illustration of the illumination
and projection optics of the 248-nm KrF excimer laser
stepper prototype developed at AT&T. [From Ref. 26]

Best Focus + 0.5 μm

Best Focus

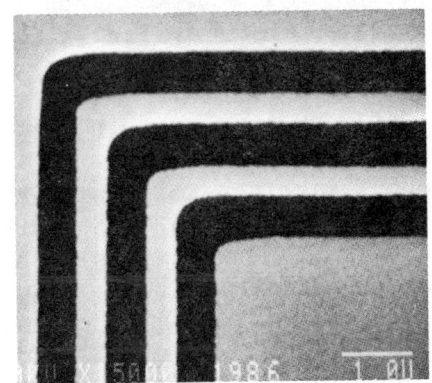

Best Focus − 0.5 μm

0.5 μm Lines and Spaces

Fig. 4.19   Images obtained with the AT&T excimer laser stepper prototype of Fig. 4.18. The patterns shown were etched into a trilevel resist structure after 248-nm exposure of a 0.5-micron-thick Shipley 2400-17 imaging layer. [From Ref. 26]

spectrally narrowed to a bandwidth of $\sim$0.004 nm. An automatic wavelength control system is incorporated to stabilize the laser center-frequency with an accuracy of 0.001 nm and also to tune it within the free-running bandwidth of 1 nm. The laser is pulsed at a repetition rate of 200 Hz and produces an average output power of 1.5 W after spectral narrowing. A temperature control system is added to compensate for atmospheric effects on image placement and focus. The overall system is illustrated schematically in Fig. 4.20. The wavelength dependence of distortion and the temperature dependence of focus for the 5:1 prototype tool are shown in Fig. 4.21. Images obtained with this machine in a new deep UV resist, Shipley XP-8843, which indicate 0.5-micron resolution, are shown in Fig. 4.22. In addition to being capable of producing good image profiles, this photoresist also has a high sensitivity, requiring an exposure dose of only 24 mJ/cm^2. We discuss this resist at greater length in Chapter 6.

A prototype 248-nm step-and-repeat machine using the KrF excimer laser has also been developed by the Matsushita Electrical Industrial Co. This 5:1 reduction system, shown in Fig. 4.23, has evolved from a first demonstration by Sasago et al. [37] and Endo et al. [45] with a quartz lens that had an NA of 0.35 and a field size of 10x10 mm^2, to an improved prototype reported by Nakagawa et al. [57] that uses a new lens, also quartz, with a 0.36 NA and a 15x15 mm^2 field. The spectral bandwidth of the excimer laser was narrowed to 0.007 nm using an intracavity etalon. With this bandwidth, it was found from simulation results (see Fig. 4.24) that, for 0.5-micron imaging with the 0.36 NA lens, the modulation transfer function did not degrade appreciably from its maximum value for ideal monochromatic illumination. The line-narrowed output of the laser was measured to be 40 mJ/pulse upon exiting the laser and 10 mJ/cm^2 per pulse at the wafer surface. Figure 4.25 shows 0.5-micron line-space patterns exposed in Shipley/AZ 2400 resist using this prototype tool. Note that, again, due to the high absorbance of AZ 2400 in the deep UV, the image walls are highly sloped.

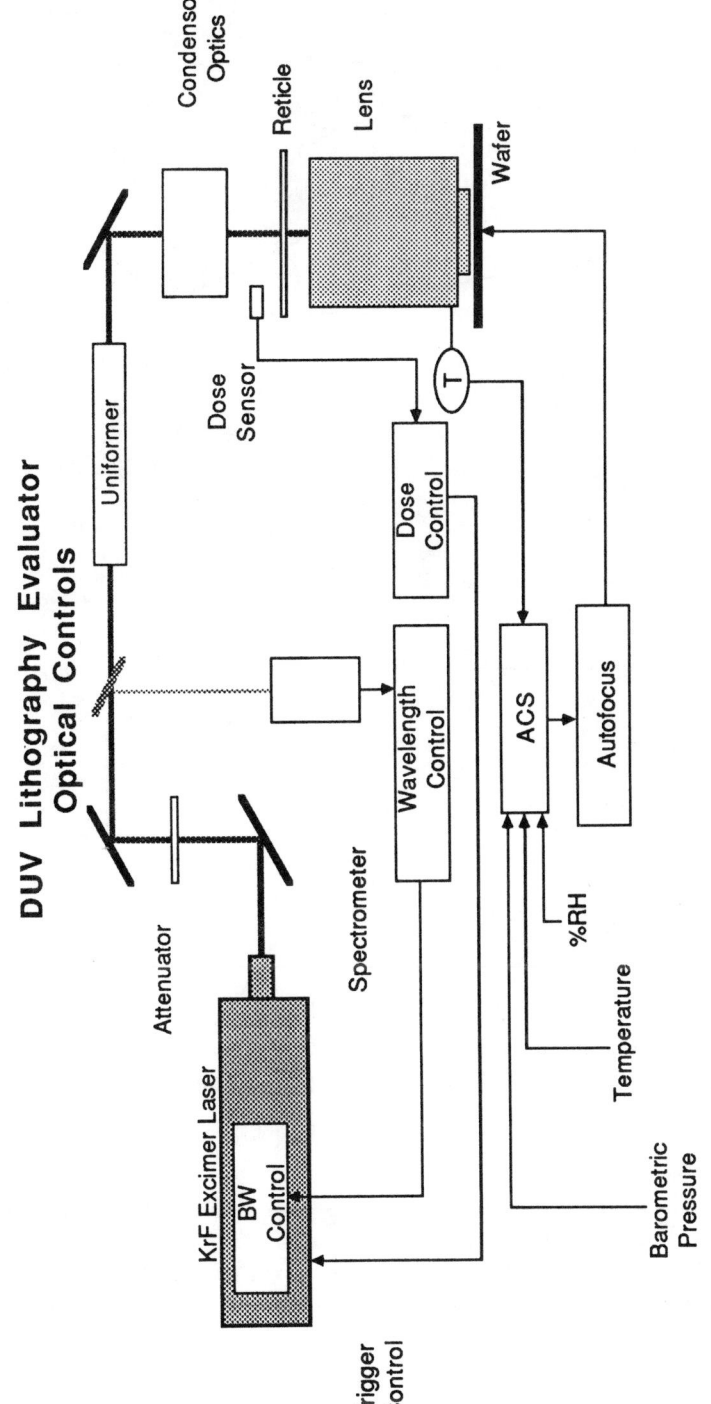

Fig. 4.20   Schematic illustration of a 248-nm excimer laser step-and-repeat prototype system developed at GCA showing various subsystems. [From Ref. 51]

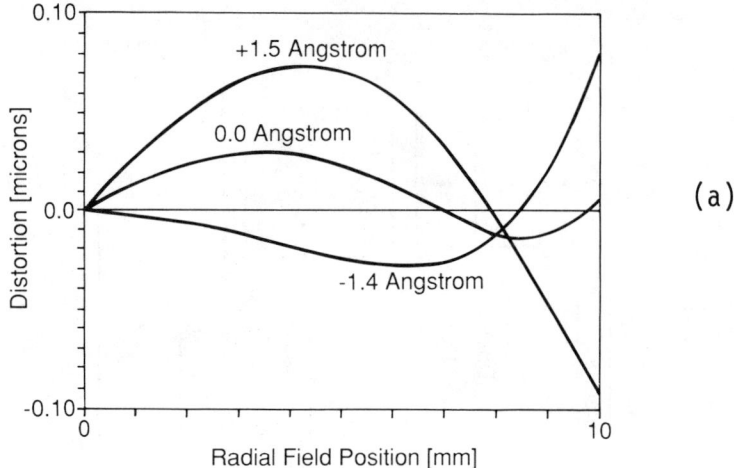

(a)

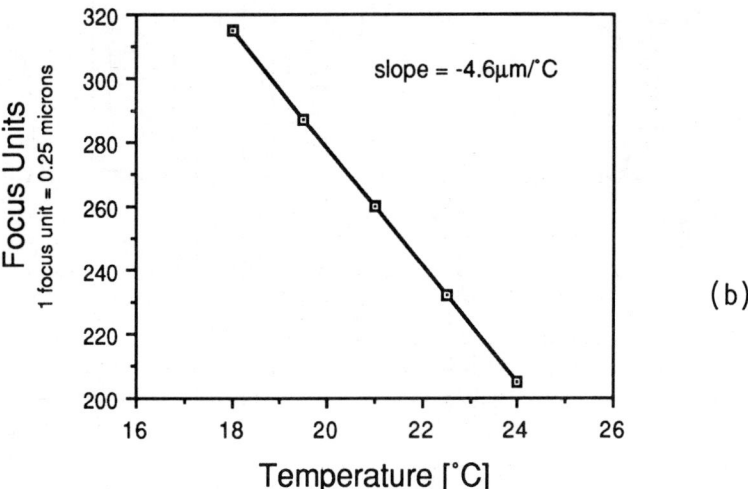

(b)

Fig. 4.21   Performance   of the GCA   248-nm   excimer
laser stepper prototype of   Fig. 4.20: (a) effect of
shift in illumination   wavelength on distortion   and
(b) dependence of focus on variation in temperature.
[From Ref. 51]

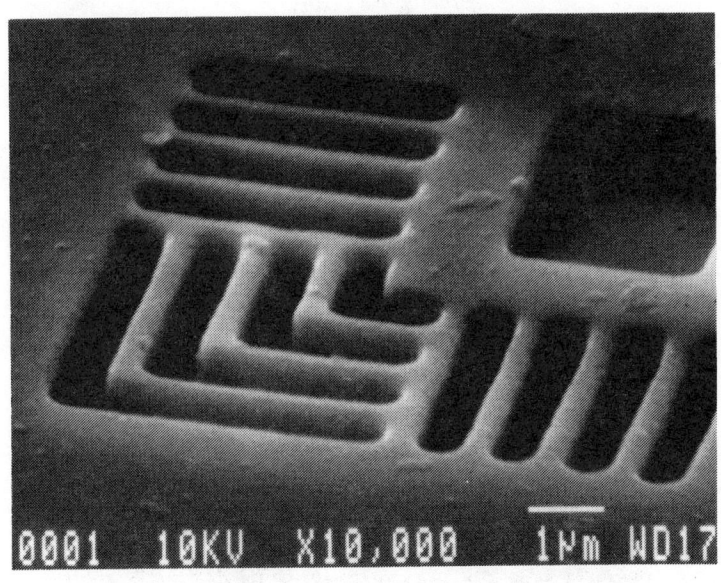

Fig. 4.22 0.5-micron images obtained with the GCA
248-nm excimer laser prototype stepper of Fig. 4.20 in
1.0-micron-thick Shipley XP-8843 resist. The chemical-
amplification resist was given an exposure dose of 36
mJ/cm^2, followed by a postexposure bake at 120 °C for
1 min. [From Ref. 65]

Fig. 4.23   248-nm KrF excimer laser stepper prototype
system developed at Matsushita Electrical Industrial
Co. [From Ref. 57]

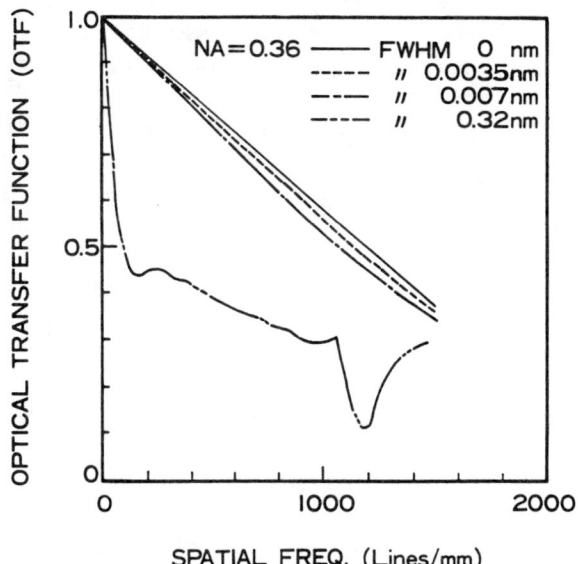

Fig. 4.24 Performance of a 0.36-NA Matsushita excimer laser prototype stepper: simulation results showing dependence of the optical transfer function versus spatial frequency curve on the illumination bandwidth (FWHM). [From Ref. 57]

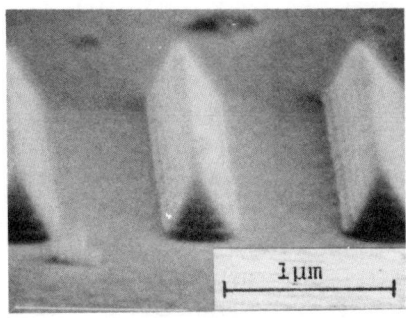

Fig. 4.25 0.5-micron-wide lines and spaces in 0.5-micron-thick Shipley MP 2400 photoresist imaged on a 0.36-NA Matsushita excimer laser prototype stepper. [From Ref. 57]

In the last few years, development of a 248-nm
KrF excimer laser stepper has also taken place at Nikon
Corporation. Their first experimental system, reported
by Kameyama and Ushida [46], had an achromatic lens
containing quartz and calcium fluoride elements that
was designed for a non-narrowed laser bandwidth of ~0.4
nm. The lens had a reduction ratio of 10:1, a numerical
aperture of 0.37 and an exposure field size of 5x5 mm^2.
Images produced in PMMA and consisting of 0.55-micron-
wide lines and spaces are shown in Fig. 4.26. Thus,
although this system was able to resolve patterns in
the 0.5-micron region, it was found that due to the
small difference between the dispersions of quartz and
CaF$_2$, adequate chromatic correction was not possible
over a larger field. Difficulties were also encountered
in cementing the lens elements together.

More recently, a new excimer laser stepper with a
monochromatic lens design has been developed at Nikon.
The new prototype has been built by making the required
modifications in one of Nikon's existing, conventional
g-line (436 nm) machines. This new system, reported by
Tanimoto et al. [68], is also designed for the 248-nm
KrF laser wavelength and has an all-quartz lens with a
reduction ratio of 5:1, a numerical aperture of 0.42, a
nominal resolution of 0.5 micron, and an exposure field
size of 15x15 mm^2. Figure 4.27 shows a schematic of the
system. The spectrally narrowed excimer laser has a
bandwidth of 0.003 nm, a center-frequency stability of
±0.001 nm, and an average power output of ~2 W at a
pulse repetition rate of 200 Hz. Uniformization of the
beam intensity to within $\pm2.5\%$ is achieved by the use
of a fly's-eye lens array and a scanning mirror. An
illumination level of ~40 mW/cm^2 is obtained in the
wafer plane. A dose controller continually measures the
integrated dose and delivers the required number of
pulses until the desired exposure level is reached. The
system claims a dose control accuracy of $\pm1.5\%$ for
doses greater than 5 mJ/cm^2. Figure 4.28 shows 0.5-
micron patterns fabricated in McDermid PR-1024MB
photoresist using the above stepper. This resist, like
Shipley/AZ 2400, also produces sloped profiles due
to excessive absorption at 248 nm.

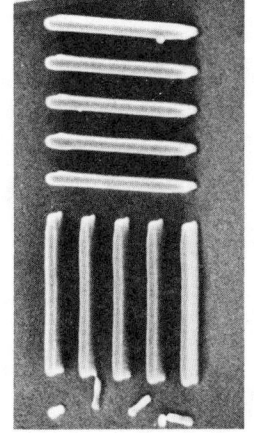
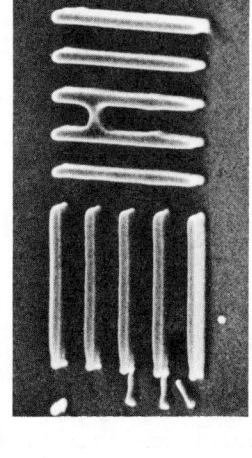
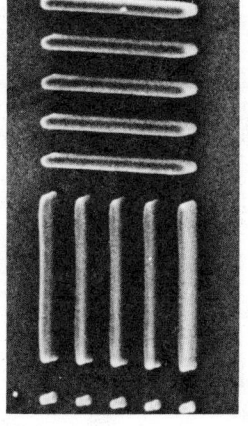
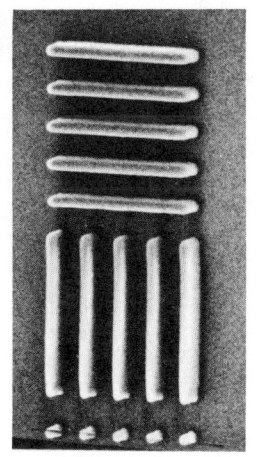
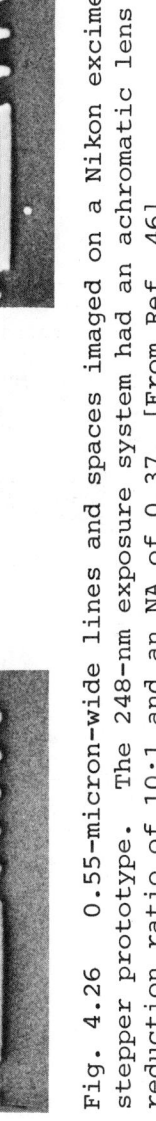
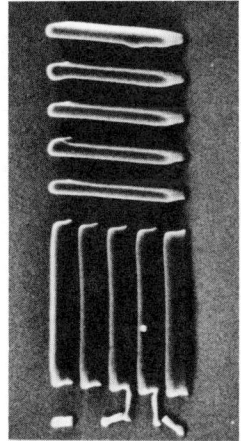

Fig. 4.26 0.55-micron-wide lines and spaces imaged on a Nikon excimer laser stepper prototype. The 248-nm exposure system had an achromatic lens with a reduction ratio of 10:1 and an NA of 0.37. [From Ref. 46]

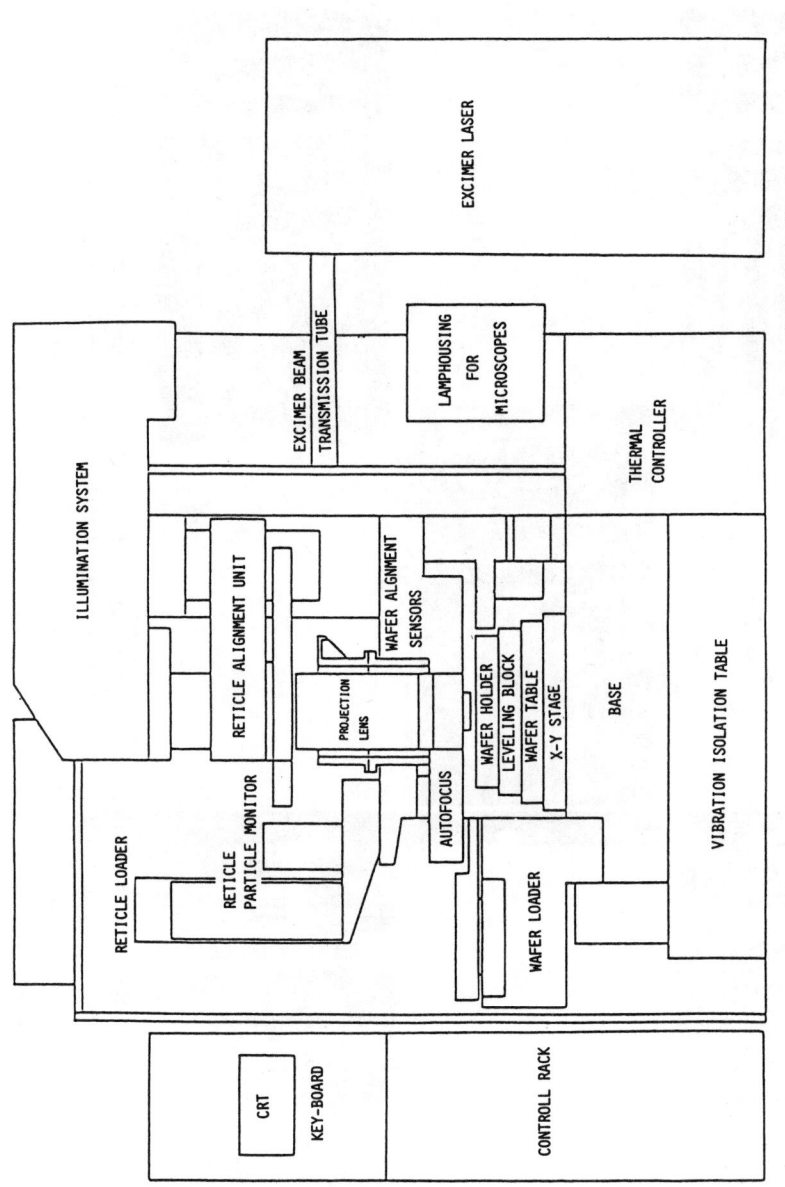

Fig. 4.27   Schematic illustration of a Nikon 248-nm KrF excimer laser stepper prototype. The system had a monochromatic projection lens with a 5:1 reduction ratio, a 0.42 NA, and a 15 mm x 15 mm exposure field size. [From Ref. 68]

FOCUS

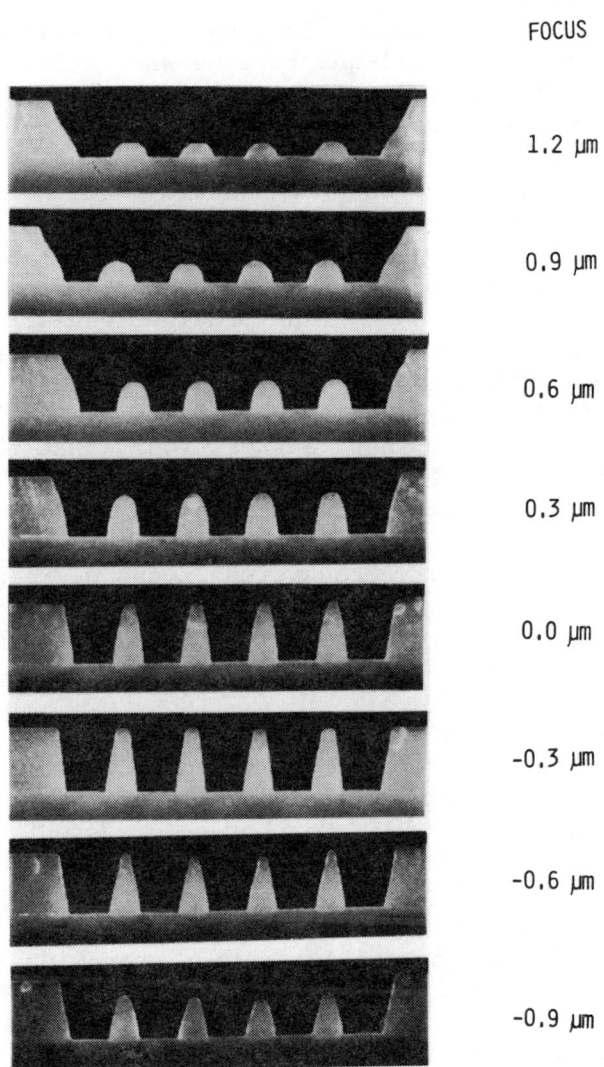

1.2 μm

0.9 μm

0.6 μm

0.3 μm

0.0 μm

-0.3 μm

-0.6 μm

-0.9 μm

Fig. 4.28 Images with 0.5-micron lines and spaces printed in McDermid PR-1024MB photoresist with the Nikon excimer laser stepper prototype of Fig. 4.27. [From Ref. 68]

Researchers at NTT have built an experimental KrF excimer laser stepper to study the effects of temporal and spatial coherence on imaging characteristics [55]. This system had an all-quartz lens with a 5:1 reduction ratio, a 0.42 NA, and a 10x10 mm^2 field size. The laser was spectrally narrowed using intracavity etalons. Different degrees of temporal coherence of the laser were obtained by varying the bandwidth of the laser in the range of 0.003-0.007 nm by adjusting the etalons. A limiting aperture placed within the laser cavity was adjusted to produce different numbers of oscillating transverse modes, which thus provided variation in the spatial coherence of the illumination. The effects of the above variations in temporal and spatial coherence were studied through the fringe contrast in a two-beam interference model. These authors found that for laser linewidth values less than 1 nm, the degree of temporal coherence had no effect on the contrast. On the other hand, the fringe visibility increased significantly when the number of transverse modes decreased below 30, indicating that lack of spatial coherence is the key factor in obtaining good image contrast in excimer laser lithography.

The effect of laser spectral linewidth on the imaging characteristics of a monochromatic all-quartz lens has also been investigated by Kajiyama et al. in simulation studies [58]. For three different values of the lens NA - 0.30, 0.35, and 0.40 - these authors calculated the image intensity profiles of a three-bar resolution target as well as the modulation transfer function for various illumination bandwidths ranging between 0.003 and 0.01 nm. Their results, reproduced in Figs. 4.29 and 4.30, show the image improvements obtained as the linewidth is narrowed. We further discuss various aspects of the coherence properties of excimer lasers and the role they play in lithography in Chapter 5.

A different approach to developing an excimer laser step-and-repeat tool has been taken by Goodall et al. in a joint program between Rutherford Appleton Laboratories and Ultratech [59]. The basic projection

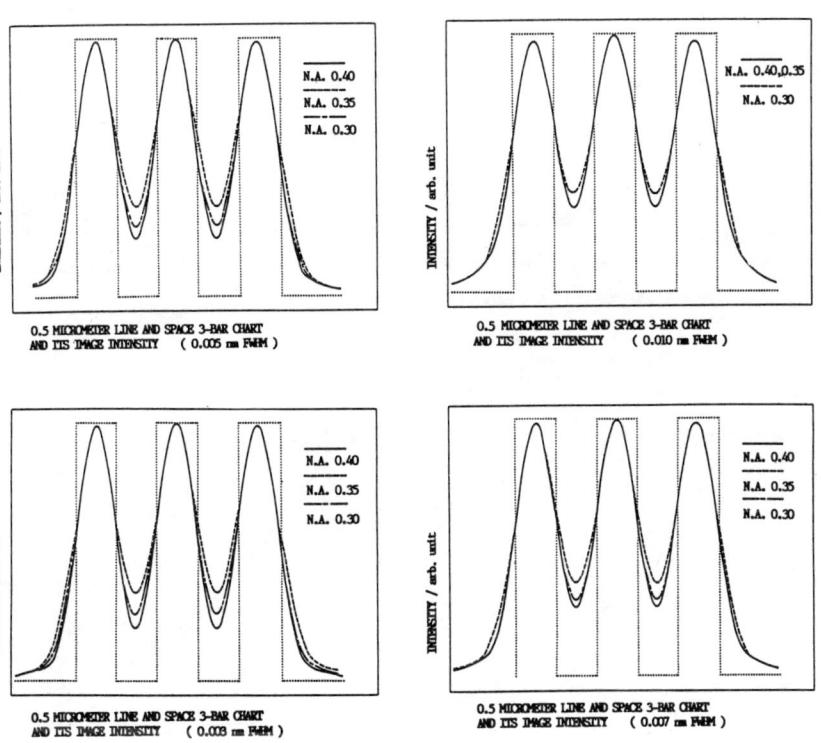

Fig. 4.29   Effect of illumination   spectral   bandwidth on the performance of a   248-nm monochromatic   quartz lens: resolution   of a three-bar target for   different bandwidths for three values of lens NA. [From Ref. 58]

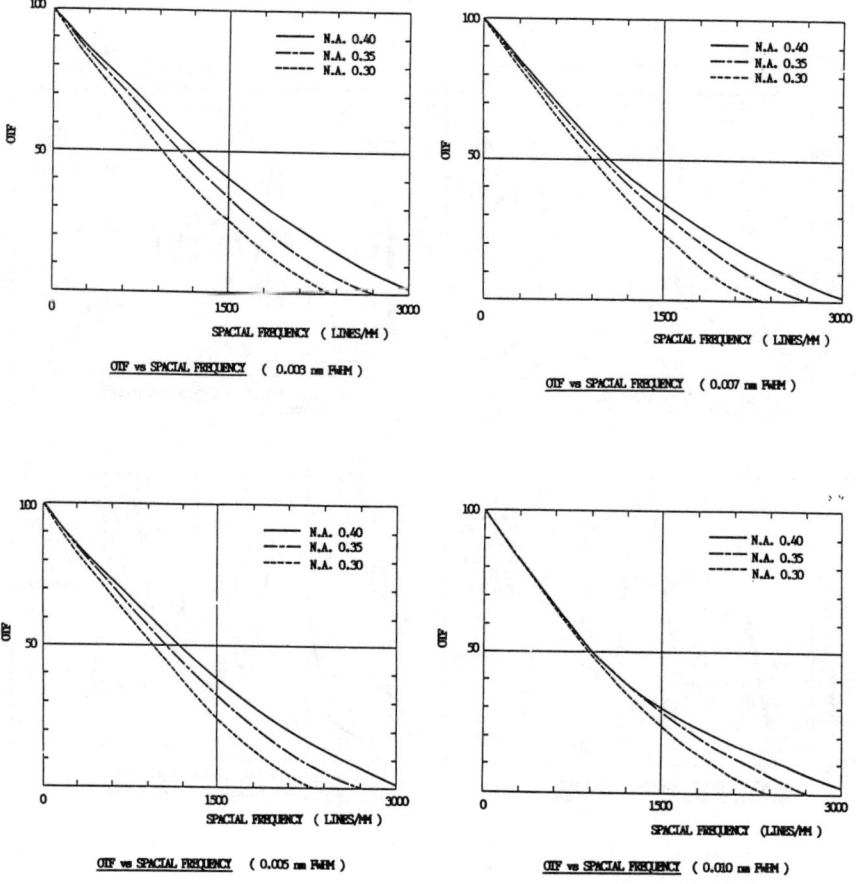

Fig. 4.30   Effect of   illumination   spectral bandwidth
on performance of a 248-nm monochromatic quartz lens:
optical   transfer function versus spatial frequency for
three values of lens NA. [From Ref. 58]

system chosen in this effort is the Wynne-Dyson design, which is a catadioptric system with a 1:1 magnification [100,101]. This choice was motivated by the observation that reflective and catadioptric systems offer certain advantages over all-refractive, all-quartz lenses that require line-narrowing of the excimer laser. The Wynne-Dyson catadioptric design uses a mirror as the main power of the system, permitting chromatic correction over the full free-running bandwidth (~ 0.4 nm) of the excimer laser with the addition of a two- or three-element achromatic lens. The monocentric design, with object and image planes on opposite sides of the center of curvature, also produces image correction over a larger field size. A schematic of the above system, designed for the KrF 248-nm wavelength, is shown in Fig. 4.31. Its refractive elements are made from fused silica and lithium fluoride; in actual construction, the LiF element was split into two 45° prisms in order to separate the object and image planes. The system had a numerical aperture of 0.30 and its exposure field was a 60-mm-diameter hemicircle that could accommodate a 21x21-mm^2 square or a 40x15-mm^2 rectangle. A non-line-narrowed excimer laser was used with a beam uniformizer to obtain an exposure level of ~ 1 mJ/cm^2 per pulse. Lithographic experiments were performed in Shipley/AZ 2400 resist to demonstrate a resolution capability in the vicinity of 0.6-0.7 micron. An improved design with a larger NA should make it possible to deliver 0.5-micron performance.

To summarize, in this section we have discussed the large number of efforts under way and the enormous progress made in realizing practical submicron excimer laser step-and-repeat projection tools. We can readily see, from the results already obtained, that production microlithography in the half-micron region with excimer laser steppers is a near certainty. With additional advances in optical systems and improvements in resist materials, we may expect that the technology of excimer laser lithography will be a strong candidate for sub-half-micron device fabrication as well.

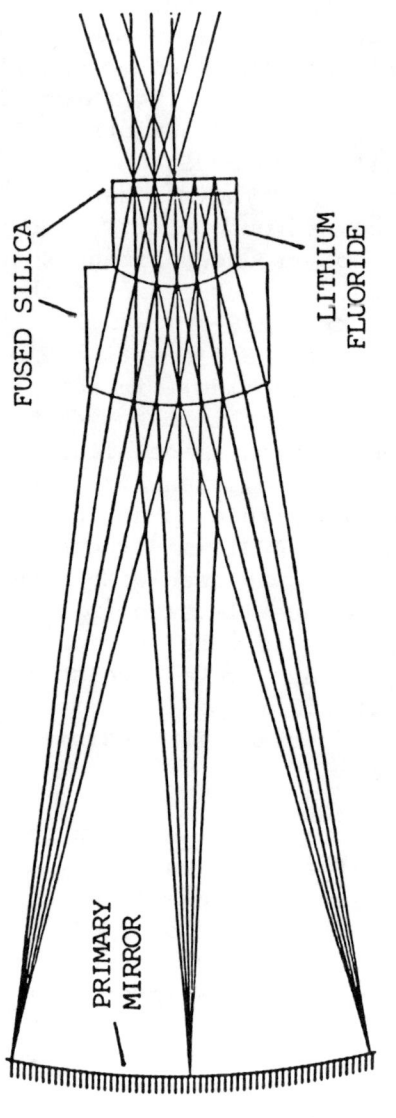

Fig. 4.31   Schematic of a large-field Wynne-Dyson-type catadioptric system for projection imaging at 248 nm. In addition to a primary mirror, the achromatic design uses quartz and LiF refractive elements and can accommodate the ~ 0.5-nm bandwidth of a free-running excimer laser. [From Ref. 59]

4.2.3. Excimer Laser Pattern Generation

Another application related to semiconductor chip manufacturing in which projection lithography using an excimer laser has been employed to advantage is artwork generation for mask fabrication. The key feature of the laser that renders it attractive in this case is its high power, which makes it possible to cut down the plate exposure time significantly. A practical excimer laser mask maker has been demonstrated by Hafner [34, 60]. In this apparatus, a commercial GCA Model 3600 pattern generator was modified for use with a 308-nm XeCl excimer laser. A schematic of the prototype is shown in Fig. 4.32. The laser output, typically 100 mJ/pulse at 200 Hz, was coupled into the condenser unit through a fiber optic cable and exposures were made through a new, 20:1 fused silica objective lens. The lens had a numerical aperture of 0.15, a resolution of 2 microns, and a maximum field size of 1.5x1.5 mm^2. Note that a 2-micron resolution is adequate for fabricating masks to be used with reduction steppers for submicron wafer lithography. A variable aperture located in the condenser and consisting of two pairs of straight edges was used, along with the x and y table movements, for pattern generation. Figure 4.33 shows how patterns of various shapes and lines of different widths are generated with such a dual aperture. The laser power output was continuously monitored and used in an active feedback loop to adjust various laser parameters to ensure uniformity of exposure. The tool was used in a fast flash-on-the-fly mode with conventional novolak-type photoresists. In reported experiments, a number of different 5:1 reticles have been generated, the most complex of which required 2x10^6 flashes from the laser, corresponding to a total running time of 27.1 h. It would have taken 270 h to generate the same reticle on a conventional GCA Model 3600 pattern generator. For the average mask, the time saved in exposure with the excimer laser artwork generator was found to be 90%, making it a far more economical mask-making tool than electron-beam machines. In addition to the above IBM effort, GCA has developed a similar excimer laser pattern generator and marketed it as a commercial unit.

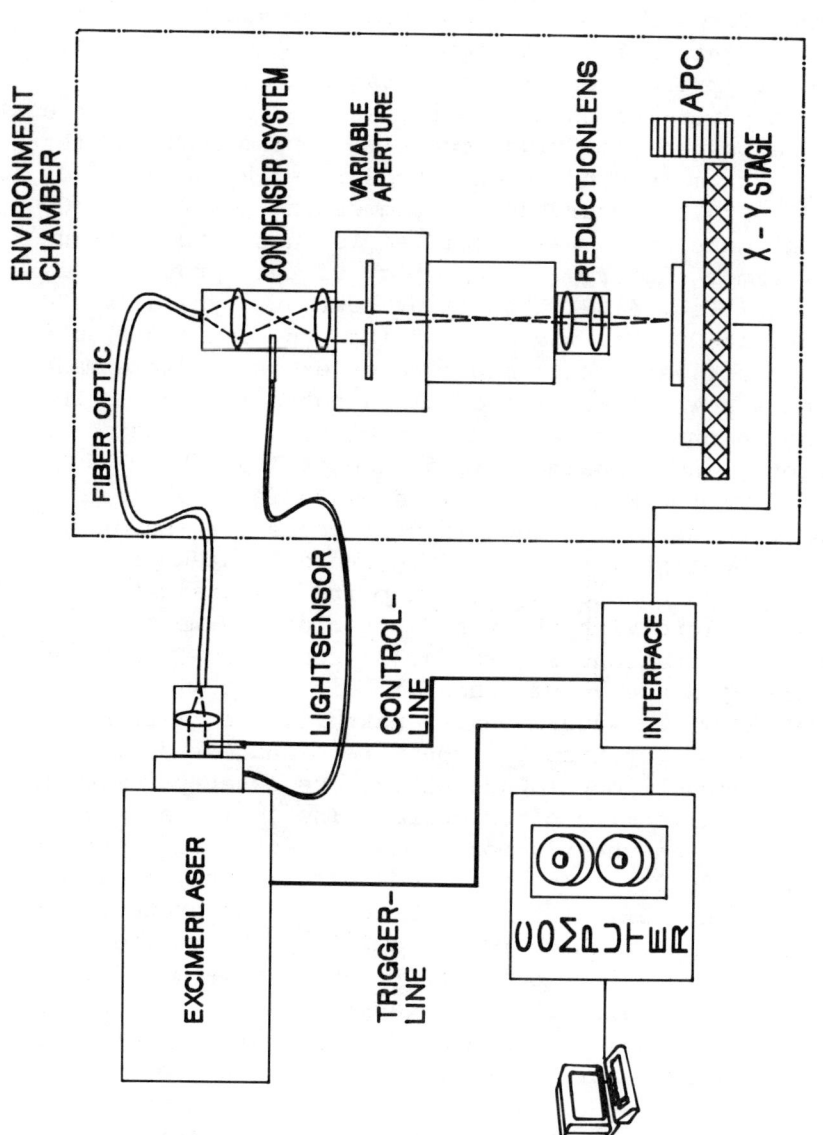

Fig. 4.32   Schematic of an excimer laser pattern generator developed at IBM.
[From Ref. 60]

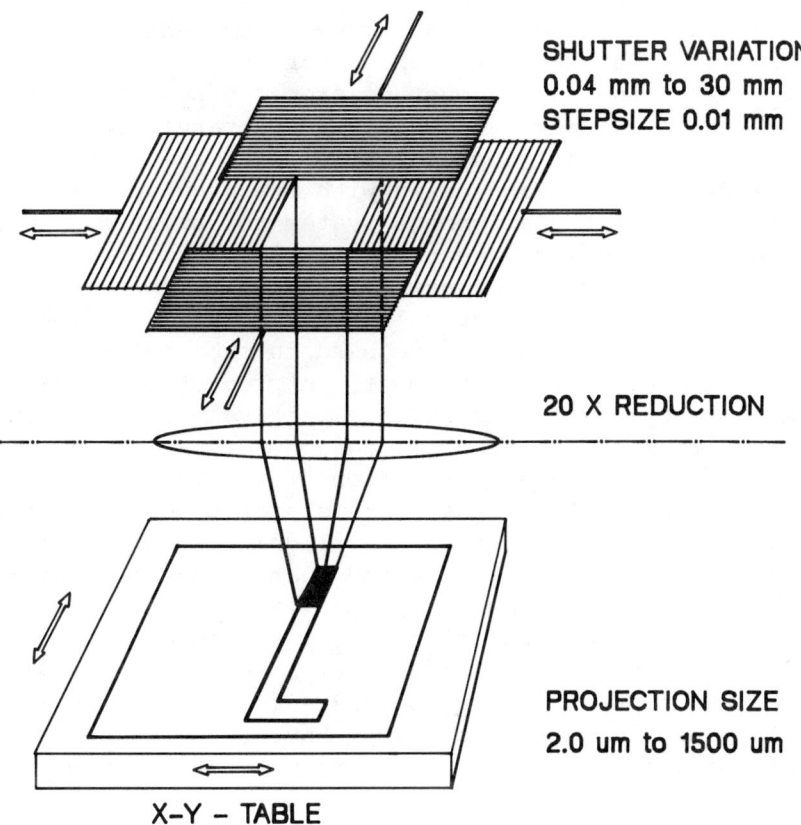

SHUTTER VARIATION
0.04 mm to 30 mm
STEPSIZE 0.01 mm

20 X REDUCTION

PROJECTION SIZE
2.0 um to 1500 um

X-Y - TABLE

Fig. 4.33   Use of a variable aperture consisting of
two pairs of straight edges to generate rectangular
patterns of various aspect ratios. [From Ref. 60]

4.3. ALIGNMENT TECHNIQUES

In addition to the illumination system and the
imaging optics, another important subsystem in the
microlithography machine is the mask-to-wafer alignment
system. Typically, a complete wafer fabrication process
cycle involves use of the lithography tool more than a
dozen times with different masks, in addition to a
myriad of other process steps. For exposure with each
new mask, the wafer must be properly positioned with
respect to the mask so that the patterns on the new
mask will overlap appropriately with the patterns on
the wafer already fabricated with the previous mask. It
is the task of the alignment system to bring the mask
and wafer into accurate mutual registration prior to
the exposure for each new mask level. The precision
of such registration typically must be a small fraction
of the minimum feature size used in fabrication of the
circuit patterns. For example, in production of 1- and
4-Mb memory chips using, respectively, 1- and 0.7-
micron geometries, the lithography system typically
must have an alignment precision of 0.3 and 0.2 micron,
respectively. Lithography systems for producing 16-Mb
devices will need to have an alignment capability of
0.15 micron, and it is expected that production of 64-
Mb chips will demand that the lithography apparatus
possess alignment precision of 0.1 micron or better. In
this section we present a brief discussion of several
alignment techniques in use today in various step-and-
repeat machines and how they have been, or could be,
modified for application in excimer laser lithography
systems.

The large variety of alignment techniques that can
be found in lithography tools have used a number of
different approaches for sensing the relative positions
of the mask and the wafer. All approaches basically
rely on detecting certain targets, fabricated on the
wafer and the mask prior to each alignment step. The
targets are illuminated by light and, through one or
more optical phenomena such as scattering, reflection,
diffraction, or interference, they produce light signals
that are then detected and analyzed to determine the

relative displacement required between the mask and the wafer to bring them into the desired alignment. A common approach found in some lithography systems, known as the bright-field alignment technique, uses mask and wafer targets that are sets of etched lines in a reflective background and arranged in the form of crosses, squares, or chevrons [102]. The targets are illuminated with visible light and the beams reflected from them are observed simultaneously in a viewing system. In the viewing microscopes the target back-grounds seem bright due to directly reflected light being collected from those areas, whereas the etched lines appear dark due to light being scattered out of the collection angles of the microscope objectives. The alignment process consists in aligning the line patterns of the pair of targets as seen in the viewing system by producing the appropriate relative displacement between the mask and the wafer using a conventional laser interferometer.

Another class of alignment approaches has used one or more diffraction gratings as alignment targets [103-105]. In these techniques, the intensity and/or phase information contained in light diffracted from a wafer grating (and in some cases, again through a mask grating) is appropriately analyzed to determine the relative mask-wafer displacement necessary to produce alignment. As an example, we describe the alignment system developed by van den Brink et al. and found in ASM g- and i-line steppers [103]. A schematic illustration of this double-diffraction technique is given in Fig. 4.34. The system consists of two identical alignment units for independent alignment along two orthogonal axes. Each unit consists of its own gratings on the wafer and the mask and the optical system of Fig. 4.34. The bundles of light rays that illuminate the wafer and mask gratings travel through the main projection lens; hence such a system is called a 'through-the-lens' (TTL) alignment system. Linearly polarized light from a He-Ne laser enters beneath the mask level and is directed by a beamsplitter through the projection lens onto the wafer grating target. The wafer target, acting as a reflection grating,

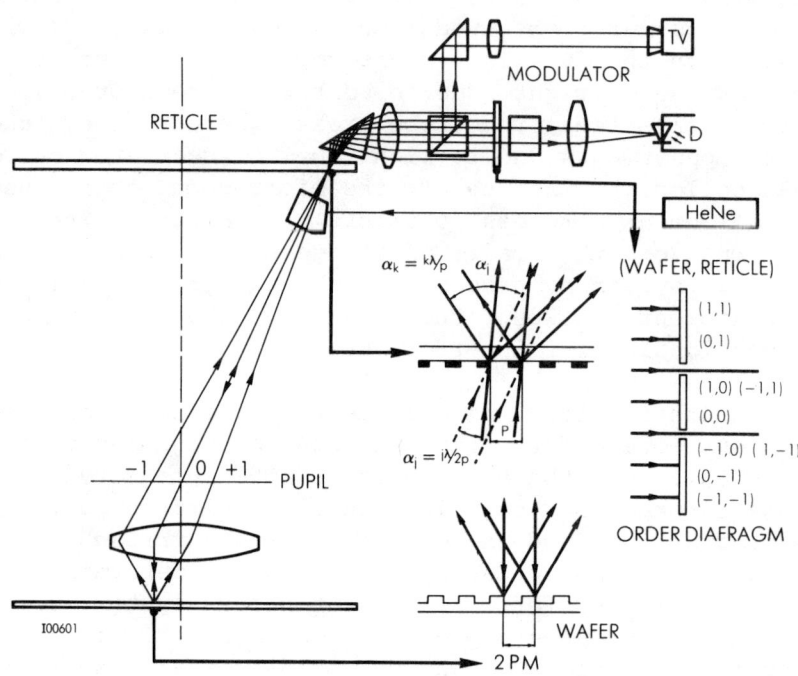

Fig. 4.34   Schematic   illustration   of   a   through-the-
lens double-diffraction alignment technique developed
at ASM for use in g- and i-line steppers. The alignment
light is provided by a 632.8-nm He-Ne laser. [From Ref.
103]

diffracts the incident beam into several diffraction orders. (The 0 and +1 orders are shown in Fig. 4.34.) Several of these diffracted beams are collected by the projection lens and imaged onto the mask grating target. Note that due to the large difference between the alignment wavelength (633 nm) and the exposure wavelength (436 or 365 nm, for g-and i-line lenses, respectively), the focus setting and magnification of the projection lens must be suitably corrected for proper imaging of the wafer target on the mask. The mask target, acting as a transmission grating, diffracts each wafer diffraction order into various mask diffraction orders. A spatial filter in the detection system selects certain specific pairs of the above double-diffracted beams. The signals are then analyzed by electronic modulation techniques to extract the wafer-position information as well as to produce the desired mask-wafer displacement necessary for alignment. An overlay precision of ± 0.15 micron (3σ) has been claimed for the most advanced production submicron steppers marketed by ASM.

A variation of the double-diffraction alignment method described above has been implemented in a 248-nm KrF laser step-and-repeat tool by Higashikawa et al. at Toshiba [49]. The projection lens in the stepper was of an achromatic design, employing quartz and calcium fluoride elements. The reduction ratio of the lens was 10:1, its numerical aperture 0.37, and its field size 5x5 mm^2. A He-Ne laser was used to provide the alignment illumination. The He-Ne beam passed through a window in the mask and the projection lens (thus, this is also a TTL system) and then illuminated a checker-patterned grating target on the wafer. The two first diffraction orders from the wafer grating were collected by the projection lens and imaged onto the mask target, which consisted of a stripe grating. The focus correction required by the projection lens for imaging with the 633-nm alignment wavelength was 52.4 mm (with respect to the 248-nm exposure wavelength), and was optically accommodated in the system by employing two folding mirrors (see Fig. 4.35) in the wafer-to-mask beam path. Interference between the mask grating and the wafer-

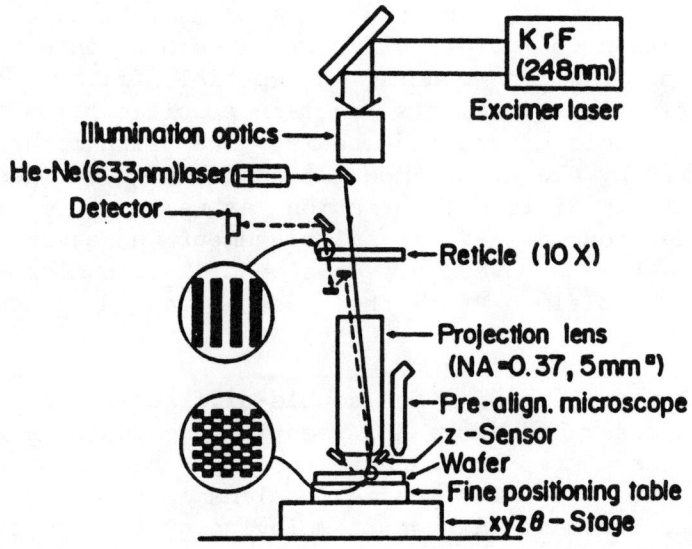

Fig. 4.35   Adaptation   of   a through-the-lens double-
diffraction alignment concept in a 248-nm KrF excimer
laser stepper with a 10:1 achromatic projection lens.
The alignment light is provided by a 632.8-nm He-Ne
laser. [From Ref. 49]

diffraction orders incident on it generated a moire pattern that changed periodically as a function of the relative mask-wafer displacement; these changes were analyzed to provide the desired alignment signal. In preliminary experiments using patterns in doped poly-silicon as alignment marks, the authors have reported achieving an overlay accuracy of $\pm$0.2 micron (3σ).

Another alignment approach found in existing tools uses what is known as the dark-field detection method. In this method, marks in the illuminated wafer target are observed not by collecting the light rays directly reflected from the background, but rather, by blocking all specular reflection and collecting only the light that is scattered by the edges of alignment marks. This is done by illuminating the target through an annular cone and collecting rays scattered into the hollow of the cone. Alternatively, the illumination may fill a certain cone angle around its axis and the detection system may collect rays scattered in an annular cone around the illumination cone. The light collected from the wafer target may then be viewed through marks on the mask to generate the desired signal corresponding to the relative mask-wafer displacement. Many different variations of the basic dark-field alignment concept described above have been implemented in lithography tools made by different manufacturers.

The dark-field system available in g- and i-line steppers offered by GCA [106] is schematically shown in Fig. 4.36(a). A small portion of the light emitted by the exposure source is collected by a fiber bundle and directed to the wafer target through a hole in a fixed mirror and through the projection lens. Since alignment is performed at the same wavelength as the exposure wavelength, no focus or magnification correction of the projection lens is required for alignment. The diameter of the hole in the fixed mirror is such that the cone of illumination is half of the exposure NA of the lens. Scattered light from the wafer is collected by the lens in an annular cone coaxial with and surrounding the illumination cone. The entrance pupil of the lens and the hole in the mirror act as the effective outer and

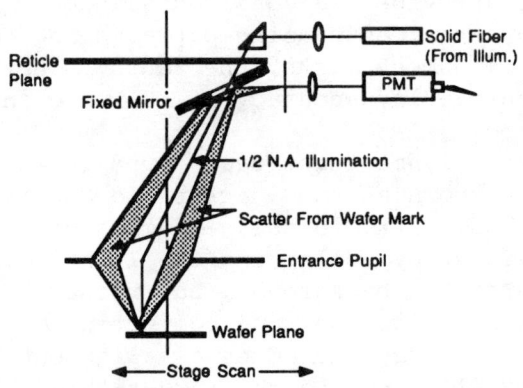

Optical Schematic

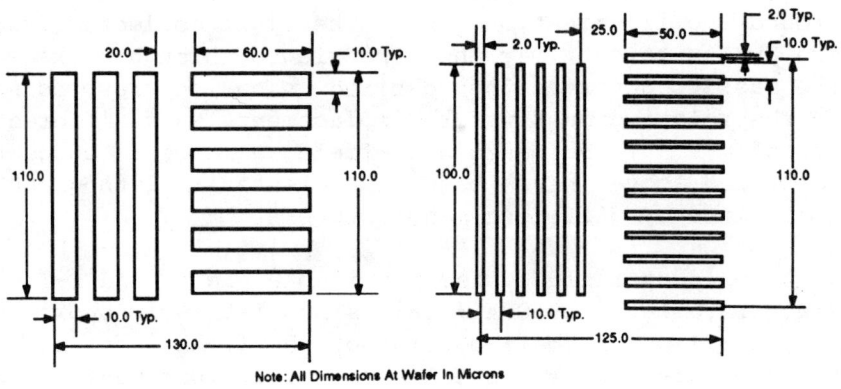

DFAS Alignment Marks

(a) Wafer Mark   (b) Reticle Window

Fig. 4.36   A dark-field, key-and-target through-the-
lens alignment system used in GCA steppers. The alignment light is at the same wavelength as that used for
the exposure, and is sampled from the mercury-lamp
exposure source. [From Ref. 106]

inner aperture stops for the scattered light. The wafer
and mask alignment targets are shown in Fig. 4.36(b).
The wafer pattern is two sets of long rectangles and
the mask pattern is a window consisting of two sets of
narrow parallel slits [see Fig. 4.36(b) for dimensions]
arranged in such a way that under perfect mask-wafer
alignment the edges of the wafer rectangles are imaged
within the mask slits. The detection system is designed
to see only those light rays that are scattered from
the edges of the wafer rectangles and then pass through
the mask slits. (Therefore, such a system is called
a 'key-and-target' dark-field alignment system.) By
optimizing the relative mask-wafer displacement so
that the detected signal is maximized, the mask and the
wafer are brought into the desired alignment. GCA has
reported achieving an overall alignment accuracy of
± 0.2 micron (3σ) in various chip fabrication process
steps.

A different dark-field alignment method has been
developed by Suzuki at Canon [107,108] and has been
available for a number of years in several types of
production lithography tools manufactured by Canon,
including contact/proximity, 1:1 full-wafer ring-field
mirror-projection scanners, and steppers. Figure 4.37
shows a schematic illustration of the basic concept.
In this technique, the mask and wafer alignment marks
are illuminated by a scanning laser beam. It is a TTL
method, in that the alignment light travels to and from
the wafer target through the main projection lens. The
scanning beam is in the form of a thin sheet produced
by cylindrical optics, with the long dimension of the
sheet being orthogonal to the scan direction. Both the
mask- and wafer-alignment marks are sets of grooves
oriented at $45°$ to the scan direction. As the laser
beam traverses across the alignment targets, light rays
scattered from the grooves are collected by alignment
microscopes and analyzed by the detection system to
generate the required signal corresponding to the mask-
wafer displacement. A noteworthy ingredient of the
above system as implemented in Canon's step-and-repeat
tools is that the alignment illumination is provided by
a helium-cadmium (He-Cd) laser. The choice of the He-Cd

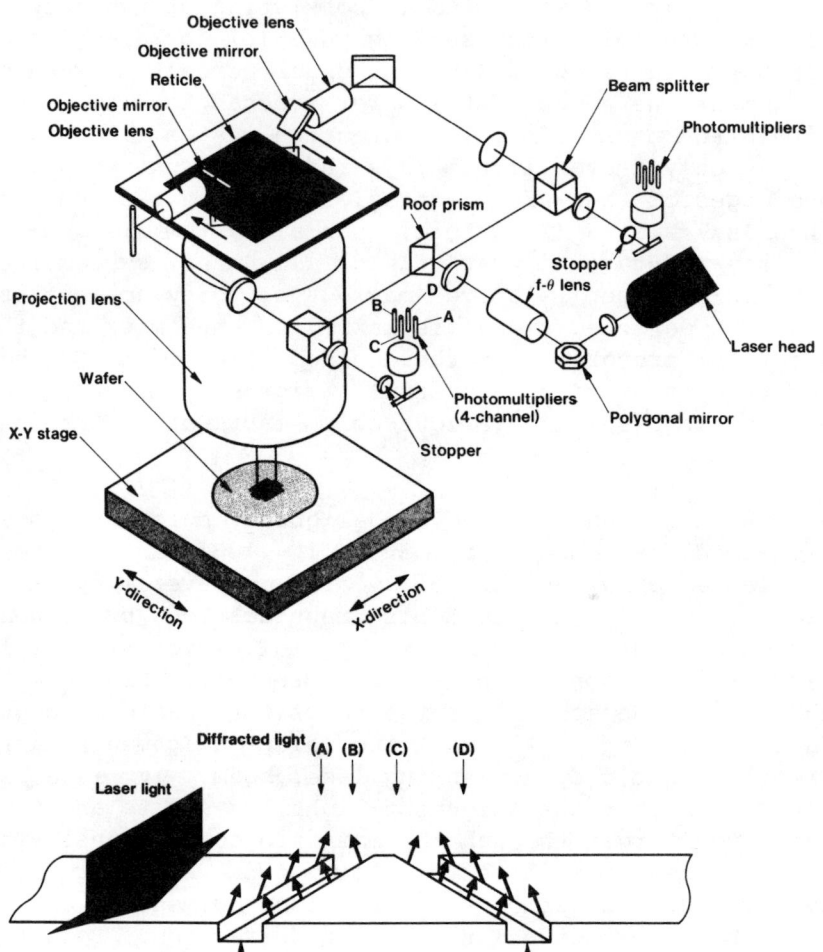

Fig. 4.37   Schematic   of   a   scanning-beam   dark-field
alignment system   developed at Canon for use in   g-line
exposure systems. The alignment is provided by a 441.6-
nm He-Cd laser. [From Ref. 107]

laser was motivated by the fact that its wavelength, 442 nm, was within the corrected bandwidth of the 436-nm g-line projection lens. This makes it possible to carry out alignment and subsequent exposure without any intervening mechanical movement. In its latest steppers employing this technology, Canon specifies an alignment precision of ±0.225 micron (3σ).

The integration of any particular alignment system into an excimer laser lithography tool will usually require certain modifications in the exposure tool; these modifications will depend on various components of the overall system. For example, if the alignment is done through the projection lens and at a wavelength different from the exposure wavelength, the optical materials used in fabrication of the lens as well as any dielectric coatings must be such that they have sufficient transmission at the alignment wavelength. In addition, adjustments must be made in the focus and magnification settings of the projection lens to allow for the (large) difference in the two wavelengths. One way to compensate for the (usually) longer alignment beam path is to introduce additional optical elements. For example, in the double-diffraction alignment system implemented by Toshiba on its 248-nm KrF excimer laser stepper described above, the beam path for the 633-nm He-Ne alignment light is corrected by inclusion of two folding mirrors (see Fig. 4.35).

Alternatively, we may consider an alignment method that relies on indirect mask-wafer referencing. Here, the alignment marks on the mask and wafer are detected separately and their positions determined and optimized independently. In such a method, signal processing and computer modelling are used to determine the position of each chip site, followed by the required translation of the wafer stage for each exposure. Such a system has been in use in various conventional steppers offered by Nikon [105, 109], for example, and has recently been implemented, without needing significant modification, on Nikon's 248-nm KrF excimer laser stepper by Tanimoto et al. [68], who have claimed achievement of an alignment precision of ±0.18 micron (3σ).

It is likely that development of alignment systems employing through-the-lens illumination of targets at the excimer laser wavelength will take place in the very near future. Such systems may use any of the basic alignment techniques discussed in this section, e.g., dark-field or double-diffraction methods. One concern in such a system would be any unwanted wafer exposure during alignment (due to the alignment wavelength being the same as the exposure wavelength). This problem may be remedied in various ways. For example, alignment may be performed at light intensity levels so much lower than those used for exposure that wafer exposure during alignment may be considered negligible. Alternatively, one may use a prealignment step to locate approximately the scribe area containing the alignment mark under the alignment beam so that device regions of the wafer are not exposed.

5. Excimer Laser Sources

In this chapter we present a detailed discussion of excimer laser sources. We describe their fundamental lasing mechanism, different excitation techniques, and various operational parameters relevant to applications in microlithography. We discuss average power output, energy per pulse, pulse repetition rate, pulse width, and spectral bandwidth. We describe these parameters in terms of both what is needed for practical lithography applications and what is available in commercially offered laser systems. In addition, we describe the recent improvements made in laser lifetime and output power stability.

Excimer lasers are a class of very efficient and powerful pulsed ultraviolet lasers that first became commercially available in the late 1970s [81,82]. The term 'excimer' is a contraction of the words 'excited' and 'dimer.' A dimer is a diatomic molecule in which the two constituent atoms are identical, e.g., Kr_2. Precisely speaking, then, a laser in which an excited state of a dimer molecule, such as Kr_2, is the lasing species is an excimer laser. The excited state of a molecule such as KrF or XeCl is more correctly known as an 'exciplex,' for 'excited complex.' Thus, most of the excimer lasers in use today in microlithography should really be called exciplex lasers, but over the years the term 'excimer' as a lasing medium has come to include all excited complexes, and we shall go along with the nomenclature.

Depending on the laser medium, excimer lasers emit light at several characteristic wavelengths from below 200 nm to above 400 nm. Some of the excimer species and the wavelengths of the intense laser transitions they produce are: ArF, 193 nm; KrF, 248 nm; XeCl, 308 nm; and XeF, 351 nm. In addition, several excimer lasers have been demonstrated in the VUV region. A list of various excimer laser transitions observed to date is given in Table 5.1, which also indicates the type of excitation scheme (see Sec. 5.3) used for each wavelength. Average

91

Table 5.1   Selected excimer laser transitions.

Lasing Excimer Species (*)	Emission Wavelength (nm)	Excitation Scheme		
		Electric Discharge	Electron Beam	E-Beam Controlled Discharge
Ar_2	126		x	
Kr_2	146		x	
F_2	*157*	*x*		
Xe_2	172		x	
ArCl	175	x		
ArF	*193*	*x*	*x*	
KrCl	*222*	*x*	*x*	
KrF	*248*	*x*	*x*	*x*
XeBr	282		x	
XeCl	*308*	*x*	*x*	*x*
XeF	*351*	*x*	*x*	*x*
XeF	488		x	

(*) Note: Lasers indicated in italic type are commercially available.

output powers of several tens of watts in the deep UV (DUV) with overall energy conversion efficiencies of >1% have already been reported. Simple and reliable systems, now commercially available, provide several options in power output, pulse energy, spectral bandwidth, and pulse repetition rate.

5.1. LASING MECHANISM

Molecules that belong to the excimer family are characterized by a bound or metastable excited state and an unstable or very weakly bound ground state. In this discussion we concern ourselves with a subset of excimer lasers known as rare-gas halide (RGH) lasers, which lase on transitions in molecules of the type RX, where R and X denote a rare-gas atom and a halogen atom, respectively. Typical potential energy curves for an RGH molecule are shown in Fig. 5.1. Population inversion in systems with such energy level schemes is readily achieved because the lower level dissociation time ($\sim 10^{-12}$ s) is much less than the upper level radiative lifetime (10^{-9} - 10^{-6} s). Excitation to the upper state RX can be produced by several mechanisms. Since the excited state is the same as the ion pair R^+X^-, recombination of the positive rare-gas ions and the negative halogen ions populates the upper level. The positive and negative ions are readily produced by collisions between high-energy electrons. Another way of producing the upper laser state involves reaction between R^* and a halogen compound; e.g.,

$$Xe^* + NF_3 \longrightarrow (XeF)^* + NF_2 .$$

5.2. EXCITATION SCHEMES

Several different approaches have been used to pump excimer laser gain media. These schemes include: (a) direct excitation by a high-energy electron beam; (b) excitation by an electric discharge controlled by an electron beam; (c) direct high-voltage electric discharge excitation; and (d) excitation by optical

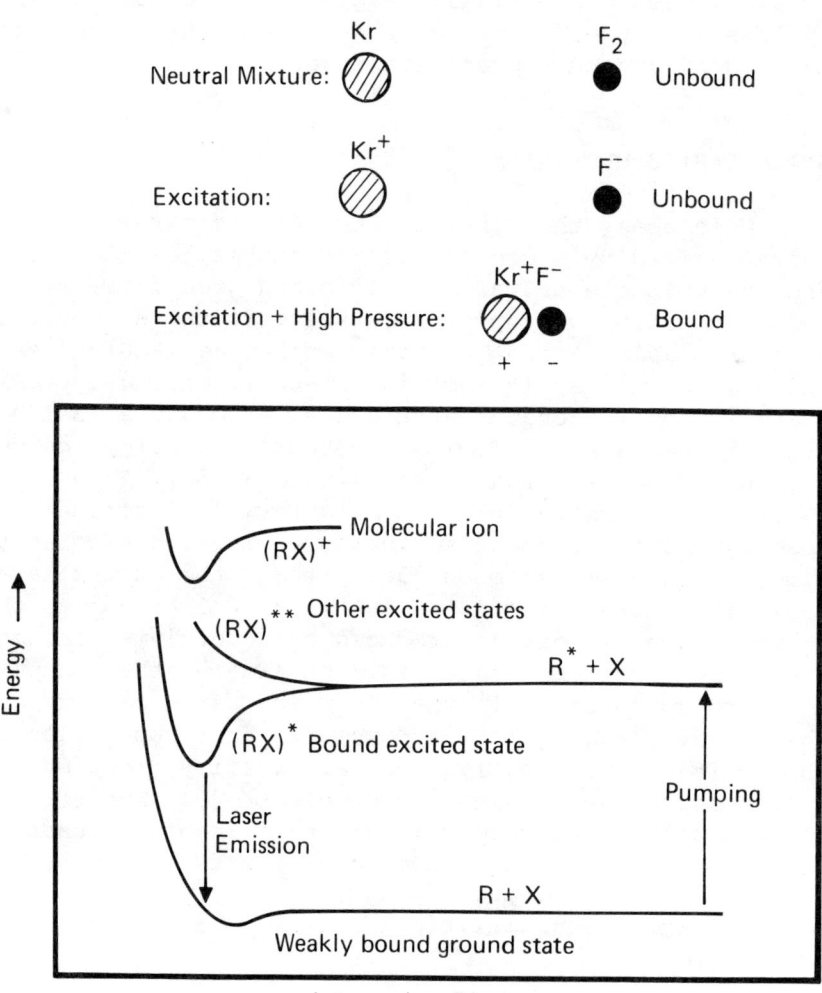

Fig. 5.1   Typical potential energy curves for rare-gas
monohalide molecules showing a laser transition from a
bound upper level to an unstable ground state.

pumping with another laser beam. Of these, the direct high-voltage electric-discharge excitation scheme is the most practical in terms of compactness and ease of operation. Typically, efficient excitation by discharge pumping requires a spatially uniform, high-voltage (20-30 kV) discharge in gas pressures of several hundreds of kilopascals (several atmospheres) with discharge gaps of a few centimeters between the electrodes. This is most easily done with a transverse discharge (see Fig. 5.2), using some sort of preionization of the gas mixture prior to the main discharge pulse. Such a transversely excited atmospheric pressure (TEA) discharge geometry, which has been widely used previously for high-power CO_2 lasers, is now found in most commercially available excimer lasers.

5.3. KEY OPERATIONAL PARAMETERS

The main parameters of excimer lasers that relate to their operation and applications in lithography and other areas are listed in Table 5.2. Observe that the parameters have been grouped in four classes. The first set, comrising average power, pulse energy, repetition rate, and pulse width, relates to delivery of the exposure dose by the excimer laser. The second group of parameters - free-running and line-narrowed bandwidths - describes the temporal coherence properties of the laser emission. The third set, which consists of beam dimensions, beam divergence, and beam uniformity, is related to the spatial characteristics of the laser including spatial coherence. Finally, the lifetime-related parameters address the maintenance and reliability aspects of excimer laser sources. We now discuss each of these parameters for various excimer lasers.

5.3.1. Exposure Dose Related Parameters

Table 5.3 contains the performance specifications of currently available state-of-the-art commercial excimer lasers. These specifications relate to the first group of parameters listed in Table 5.2 and are shown for laser operation at five wavelengths: 157 nm

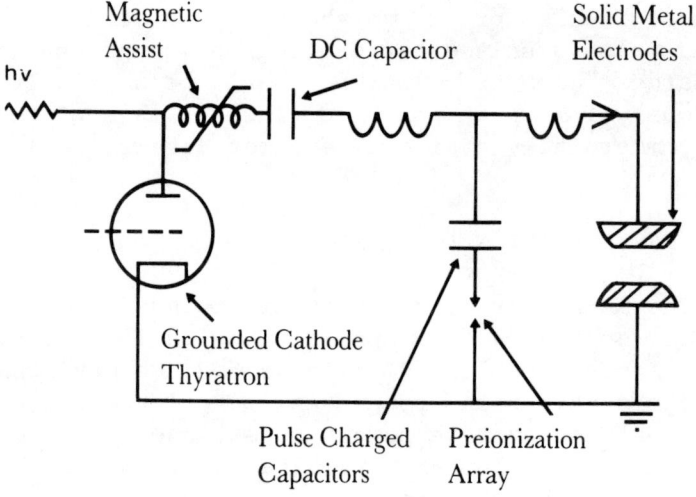

Fig. 5.2   A transversely excited atmospheric-pressure
(TEA) discharge configuration commonly used in pumping
excimer lasers.   The   high   voltage is typically 20-30
kV.   The   electrodes are typically ∼100 cm in length,
1-3 cm in width, and separated by a gap of 2-4 cm.
[Courtesy of Lumonics]

Table 5.2   Key parameters of excimer laser sources.

Exposure Dose

    o   Average Power

    o   Energy per Pulse

    o   Pulse Repetition Rate

    o   Pulse Width

Temporal Characteristics

    o   Free-Running Spectral Bandwidth

    o   Line-Narrowed Spectral Bandwidth

Spatial Characteristics

    o   Beam Dimensions

    o   Beam Divergence

    o   Beam Uniformity

Maintenance and Reliability

    o   Gas-Fill Lifetime

    o   Long-Term Lifetime

Table 5.3   Some of the key performance
parameters of excimer lasers.

	F_2	ArF	KrCl	KrF	XeCl	XeF
	157 nm	193 nm	222 nm	248 nm	308 nm	351 nm
Average Power (W)	0.05	40 (60)	4.8 (6)	100	75 (100)	
Energy per Pulse (mJ)	6	175 (300)	125	300 (1500)	200 (1500)	
Repetition Rate (Hz)	10	400	150	500	500	
Pulse Width (ns)	20	15	15	25	20	

with F_2, 193 nm with ArF, 222 nm with KrCl, 248 nm with KrF, and 308 nm with XeCl. Of these, we consider the ArF, KrF, and XeCl lasers to be of primary interest in microlithography, the power output with F_2 and KrCl operation being inadequate for practical lithography applications. Observe that each of ArF, KrF, and XeCl lasers readily produces tens of watts of average power, >200 mJ of energy in each pulse, and several hundred hertz repetition rate. In each column in Table 5.3, the values not in parentheses can be obtained simultaneously from the same physical laser unit. The parentheses indicate that a value within them is available, but the user will have to sacrifice one or more of the other parameters in the same column. For example, with KrF operation, a pulse energy of 1500 mJ is available from a laser with modified electrodes, but at lower average power and repetition rate than 100 W and 500 Hz, respectively.

5.3.2. Beam Characteristics and Spatial Coherence

Since an excimer laser usually does not have an optical resonator cavity, its beam dimensions and beam divergence are determined by the electrode geometry. Most excimer lasers have highly non-Gaussian, nearly rectangular beams with large cross sections, commonly 2-3 cm in the long dimension (usually horizontal) and 1-2 cm in the short dimension (usually vertical). The corresponding beam divergence values are found to be $\sim$6 mrad in the long dimension and $\sim$2 mrad in the short dimension. Figure 5.3 shows typical examples of the beam profiles of an excimer laser. The large beam size and divergence and the large mode volume also emphasize the almost superradiant nature of the excimer laser emission, which results from the high gain in the laser medium. These beam characteristics have a profound bearing on the spatial coherence properties of excimer lasers and the role they play in lithography.

In comparison with conventional 'coherent' lasers such as an Ar-ion or a He-Ne laser, excimer lasers have extremely poor spatial coherence and, therefore, they do not produce speckle in lithography experiments. As

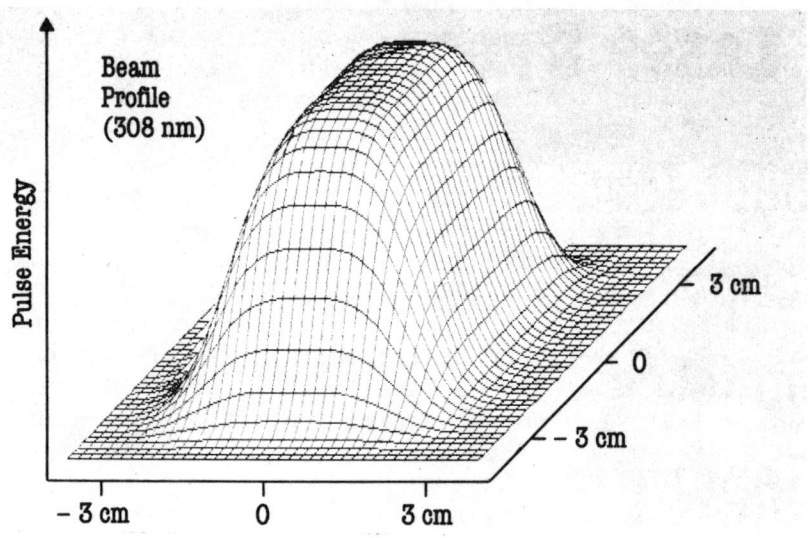

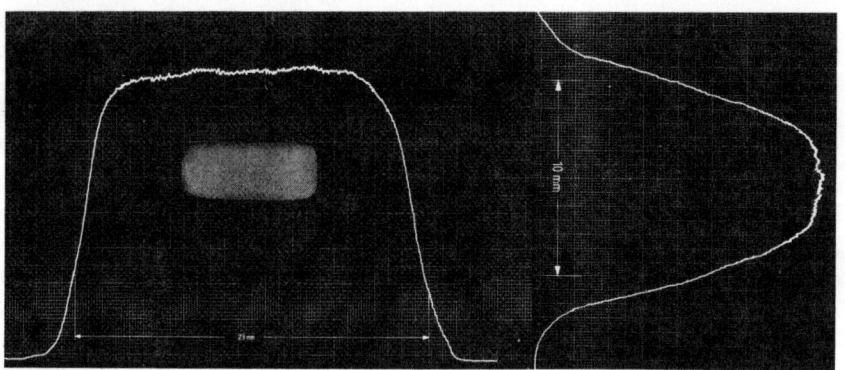

Fig. 5.3 Typical intensity profiles of the output
beam of an excimer laser. [Courtesy of Lambda Physik]

discussed previously in Sec. 2.2.1.3, speckle is the random interference pattern that is produced when, on illumination of an object with a spatially coherent wavefront, any scattering at an optical surface causes different parts of the wavefront to interfere with each other constructively and destructively. An example of a speckle pattern produced by illumination with spatially coherent light from a single-transverse-mode laser is shown in Fig. 5.4. The value of contrast in such a speckle pattern is given by $1/\sqrt{N}$, where N is the number of independent, spatially coherent wavefronts, or modes, used to illuminate the object. Thus, in the example of Fig. 5.4, N = 1 gives a contrast of 1. As N increases, the speckle contrast decreases. This may be observed in Fig. 3.3, which shows how the quality of the bar target images made with pulses from an Ar-ion laser improves as the number of pulses used in making the exposure increases. In illumination with an excimer laser, N may have a value of several thousands or even hundreds of thousands, which results in an interference pattern with effectively zero contrast, i.e., an exposure free of speckle. A summary of the above spatial-coherence-related characteristics comparing an excimer laser with an Ar-ion laser on the one extreme and a Hg-Xe arc lamp on the other extreme is given in Table 5.4.

5.3.3. Spectral Bandwidth and Temporal Coherence

The spectral bandwidth of the radiation emitted by a laser determines its temporal coherence which may be expressed as a coherence length l_c, defined by

$$l_c = c/\Delta f = \lambda^2/\Delta\lambda,$$

where c is the speed of light, Δf the frequency spread of the laser output, and $\Delta\lambda$ its corresponding spread in wavelength. For a free-running (i.e., not spectrally narrowed by additional frequency-selective methods) excimer laser, the lasing bandwidth is typically <1 nm, which corresponds to a coherence length of ~100 microns. In comparison, the emission from an Ar-ion laser, for example, is confined within a bandwidth of

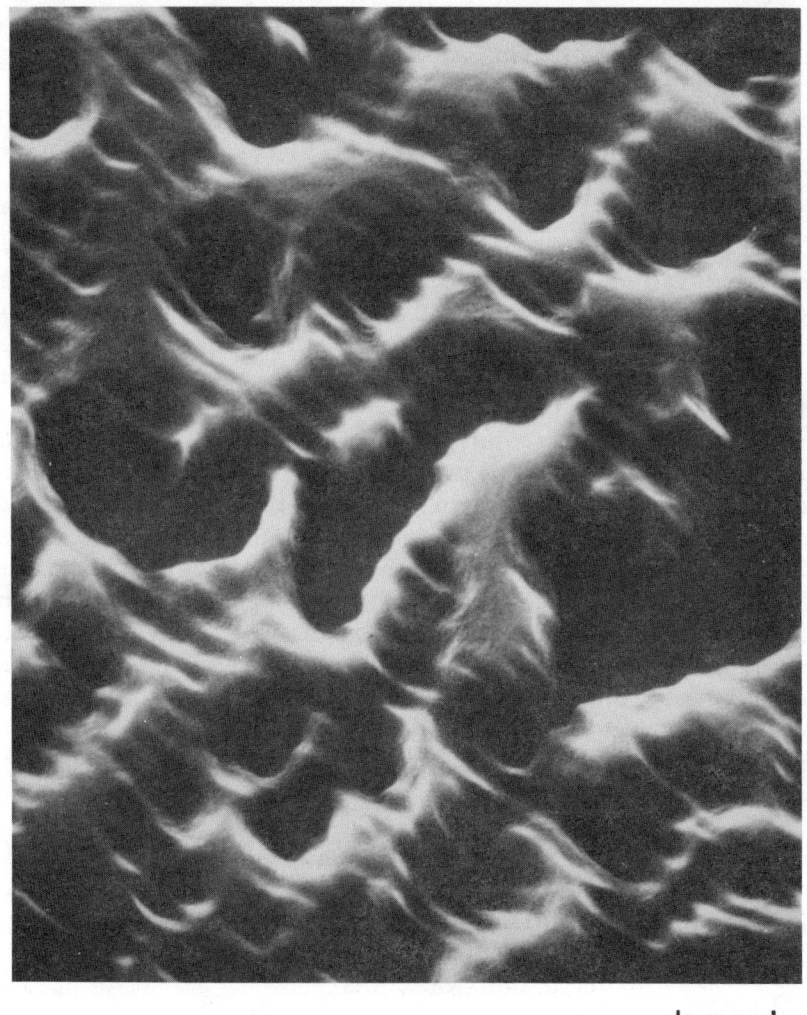

5 μm

Fig. 5.4   Speckle pattern produced when light from a
coherent source is scattered from an optical surface
and different parts of the wavefront interfere with
each other. [Courtesy of M. D. Levenson]

Table 5.4   Comparison of temporal and spatial coherence properties of a free-running excimer laser with an Ar-ion laser and a Hg-Xe arc lamp.

		Argon Ion Laser	Free-Running Excimer Laser	Hg-Xe Arc Lamp
Temporal Coherence	Spectral Bandwidth	<0.0001 nm	1 nm	10 nm
	Coherence Length	>1 m	100 μm	10 μm
Spatial Coherence	Single Transverse Mode?	Yes	No	No
	Number of Eqvt. Modes	1	10^3 -10^5	10^8
	Speckle Contrast	1	10^{-2} -10^{-3}	10^{-4}

<0.0001 nm, which produces a coherence length of several meters. On the other hand, light from a common Hg-Xe lamp used in conventional projection lithography tools may be collected in a wavelength band ∼10-nm wide, limiting its coherence length to ∼10 microns. The above comparative figures are included in Table 5.4 along with the figures relating to spatial properties.

In many excimer laser applications in lithography, the temporal characteristics of a free-running excimer laser are satisfactory. These applications include: (a) contact and proximity tools, which, for all practical purposes, are 'bandwidth-blind' since no optical image formation takes place in them; (b) full-wafer scanning projection printing systems, which have a wide working bandwidth due to their all-mirror projection system; (c) step-and-repeat tools that use achromatic reduction projection lenses designed with two optical materials, e.g., quartz and a fluoride (see examples in Table 4.1); these can be designed with color correction over the free-running excimer laser bandwidth; and (d) 1:1 imaging systems based on the catadioptric Wynne-Dyson design, in which color correction over a bandwidth of several angstroms is readily achieved as a result of the dominant imaging element being reflective.

However, in the lithography system application of excimer lasers that is currently being most extensively pursued, namely, a stepper with a monochromatic, all-quartz lens, there is a requirement for a very narrow laser bandwidth, typically 0.003 nm. For implementation in these systems, thus, the free-running excimer laser must be line-narrowed by ∼2 orders of magnitude (see Fig. 5.5) by employing certain frequency-selective devices or techniques. A method commonly used to narrow the spectral width of lasers that normally emit in a wide bandwidth is injection locking [110]. In this technique, a two-stage apparatus is used to obtain the desired narrow-bandwidth output. The first stage consists of a low-power excimer laser oscillator in which the optical cavity includes a stable resonator as well as certain intracavity frequency-tuning elements such as prisms, gratings, or etalons. In the second

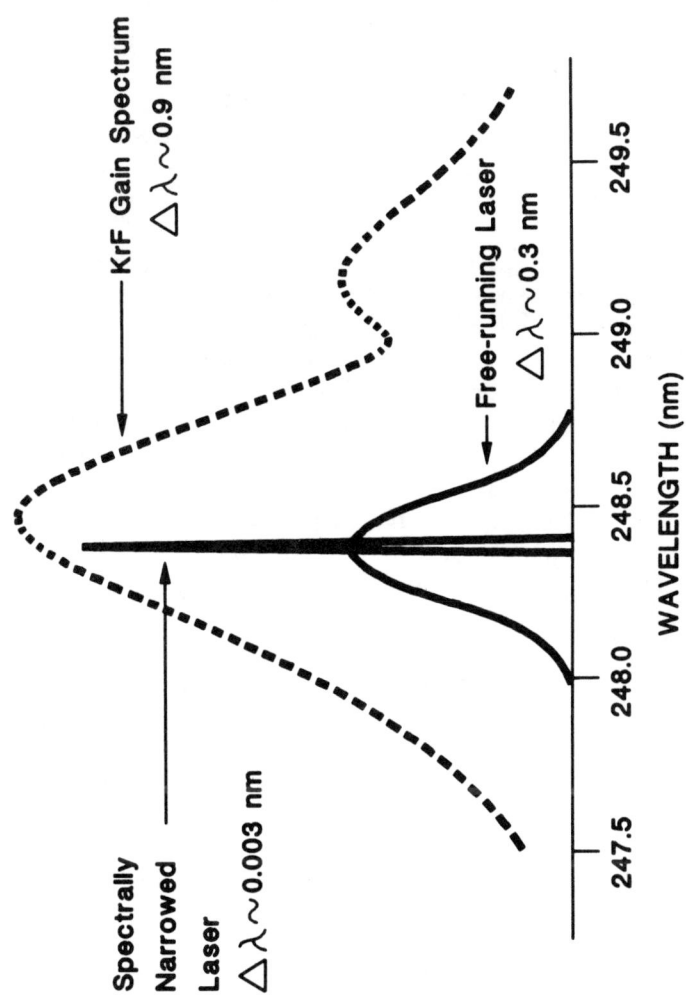

Fig. 5.5 Spectral profiles showing the gain spectrum of the KrF lasing medium, the spectral bandwidth of the free-running KrF laser, and the output bandwidth following line-narrowing. [Courtesy of Lumonics]

stage, the output of the oscillator is used to 'seed,'
i.e., initiate the stimulated emission of radiation in,
an excimer laser amplifier in which the optical cavity
consists of an unstable resonator. The second stage
provides power amplification for the input seed beam,
while keeping the narrow spectral width of the latter
essentially unchanged. Figure 5.6 shows two examples
of injection-locking configurations. In Fig. 5.6(a),
two separate laser units, each with its own discharge,
are used as the oscillator and the amplifier stages. In
the second configuration, shown in Fig. 5.6(b), the two
stages are optically two lasing units, but they share a
common discharge chamber.

In addition to providing power amplification in a
narrow bandwidth, injection locking also reduces the
divergence of the output beam, concentrating more of
its energy near the beam axis. The normal divergence of
$\sim$2-6 mrad seen in the output beam of a conventional,
free-running excimer laser is reduced by more than an
order of magnitude, to 0.15-0.3 mrad, when injection
locking takes place in an unstable resonator. This is
qualitatively illustrated in Fig. 5.7, which shows the
angular energy distribution in the output beam from a
conventional excimer laser with a stable resonator, a
laser using an unstable resonator, and an injection-
locked amplifier with an unstable resonator.

In Table 5.5 we summarize the performance of the
three primary excimer lasers, ArF, KrF, and XeCl, with
spectral narrowing. The free-running output bandwidth
is shown in the first row. The bandwidth achieved and
the average power, pulse energy, and repetition rate
available when spectral narrowing by injection locking
is employed are shown in the rest of the table. Note
that even after narrowing the line by $\sim$2 orders of
magnitude, $\sim$40% of the non-narrowed average power is
still available.

In practical implementation of a line-narrowed
excimer laser in a lithography system, the injection-
locking technique suffers from some drawbacks. The
overall injection-locked system is physically large

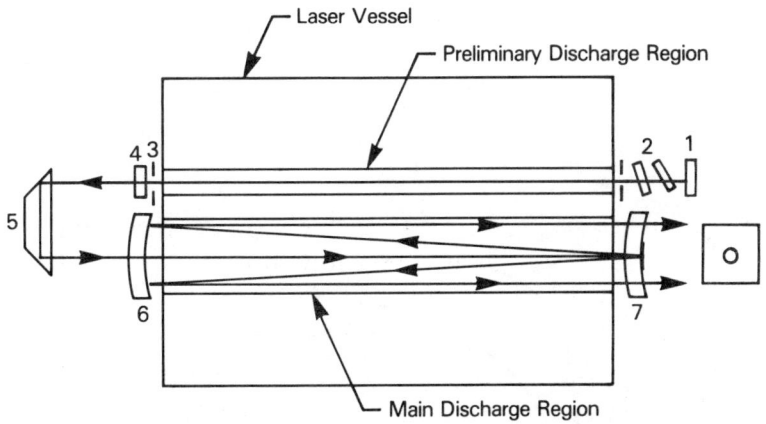

1 Master oscillator rear reflector   5 Prism
2 Etalons   6 Rear unstable resonator optic
3 Apertures   7 Front unstable resonator optic
4 Master oscillator output coupler

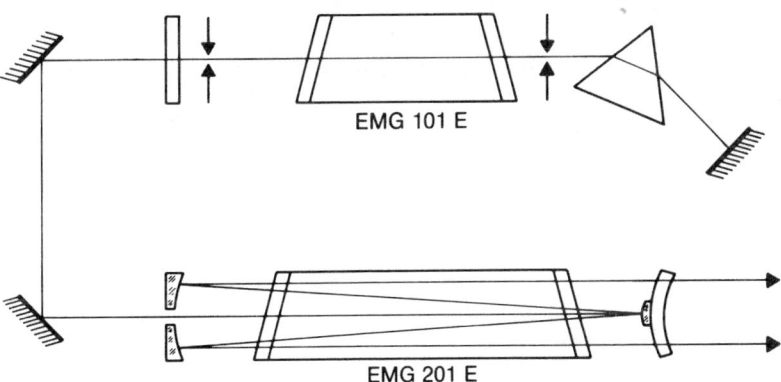

EMG 101 E

EMG 201 E

Fig. 5.6   Schematics illustrating configurations for spectrally narrowing an excimer laser by injection locking. [Courtesy of Lumonics and Lambda-Physik]

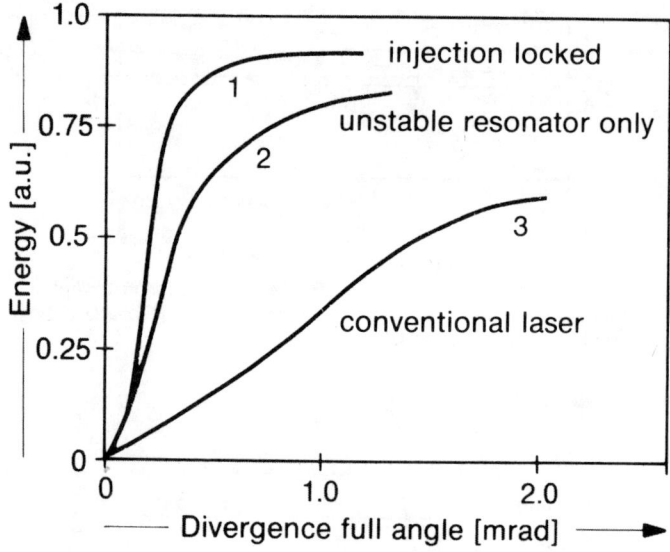

Fig. 5.7   Effect of injection locking on the output
beam divergence of an excimer laser. The curves show
the angular energy distribution in the output beam of
a conventional excimer laser with a stable resonator,
a laser using an unstable resonator, and an injection-
locked laser with an unstable resonator. [Courtesy
of Lambda Physik]

Table 5.5. Performance of excimer lasers with
   spectral narrowing by injection locking.

| | ArF | KrF | XeCl |
	193 nm	248 nm	308 nm
Free-Running Spectral Bandwidth (nm)	0.5	0.5	0.5
Operation with Spectral Narrowing			
Bandwidth with Line Narrowing (nm)	0.004	0.001	0.01
Average Power (W)	16	50	30
Energy per Pulse (mJ)	100	200	120
Repetition Rate (Hz)	250	250	250

since it basically consists of two laser units. For the
same reason, it also adds significantly to the system
expense. Further, due to the requirement of maintaining
discharge synchronization and optical alignment between
the oscillator and the amplifier, the operation and
maintenance of the laser become more complex. Finally,
injection locking, from the fundamental nature of its
mechanism, also increases the spatial coherence of the
laser radiation to an undesirable degree, making it
difficult to eliminate speckle. As a result of these
considerations, for microlithography applications it is
preferable to use a single-unit line-narrowed excimer
laser system in which the required frequency-selective
intracavity optical components have been added.

A large variety of methods exist for spectrally
narrowing an excimer laser with a stable resonator,
the difference between the various methods being the
type of optical element(s) used to restrict the lasing
bandwidth. Figure 5.8 illustrates four such techniques.
The configuration of Fig. 5.8(a) consists of a
discharge chamber (3), a front reflector (1), a
diffraction grating (5) as the frequency-selective rear
reflector, limiting apertures (2), and beam expanding
optics (4) to reduce the power density incident on
the grating. In Fig. 5.8(b), the diffraction grating
(5) is used at grazing incidence to further reduce the
incident power density, and a rear reflector (6) has
been added. The method shown in Fig. 5.8(c) employs
a pair of prisms (7) as the frequency-discriminating
optical elements, whereas in Fig. 5.8(d), a set of
etalons (8) is used to achieve the line narrowing.

Of the above, the technique employing intracavity
etalons, outlined in Fig. 5.8(d), is the most widely
used in commercial line-narrowed excimer lasers. A more
detailed illustration of this method is given in Fig.
5.9. As shown in this figure, wavelength selection is
done by a coarse etalon (E1) and a fine etalon (E2).
In addition, the center-frequency of the output beam
is continuously monitored by a monitor etalon (E3),
from which a control signal, when fed back to the fine
etalon (E2), optimizes its orientation so as to keep

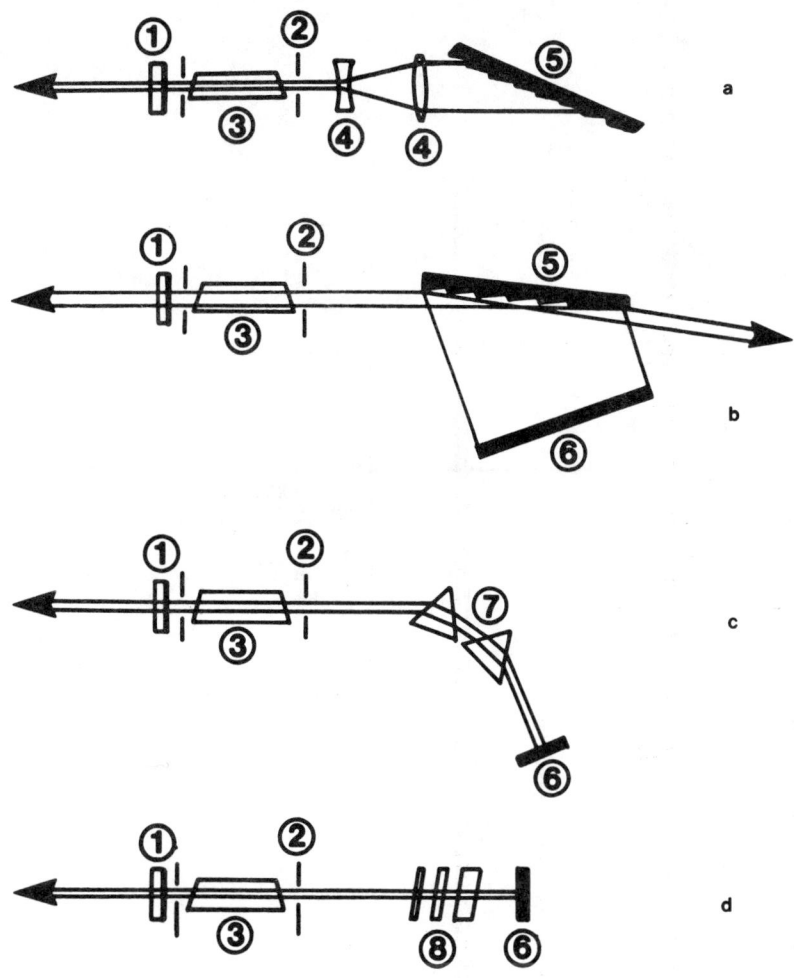

1, front reflector; 2, limiting apertures; 3, discharge
chamber; 4, beam expanding optics; 5, diffraction
grating; 6, rear reflector; 7, prisms; 8, etalons.

Fig. 5.8   Techniques for spectral bandwidth narrowing
using: (a) a diffraction grating with a beam expander,
(b) a grating at grazing incidence, (c) prisms, and (d)
etalons.   [Courtesy of T. Znotins]

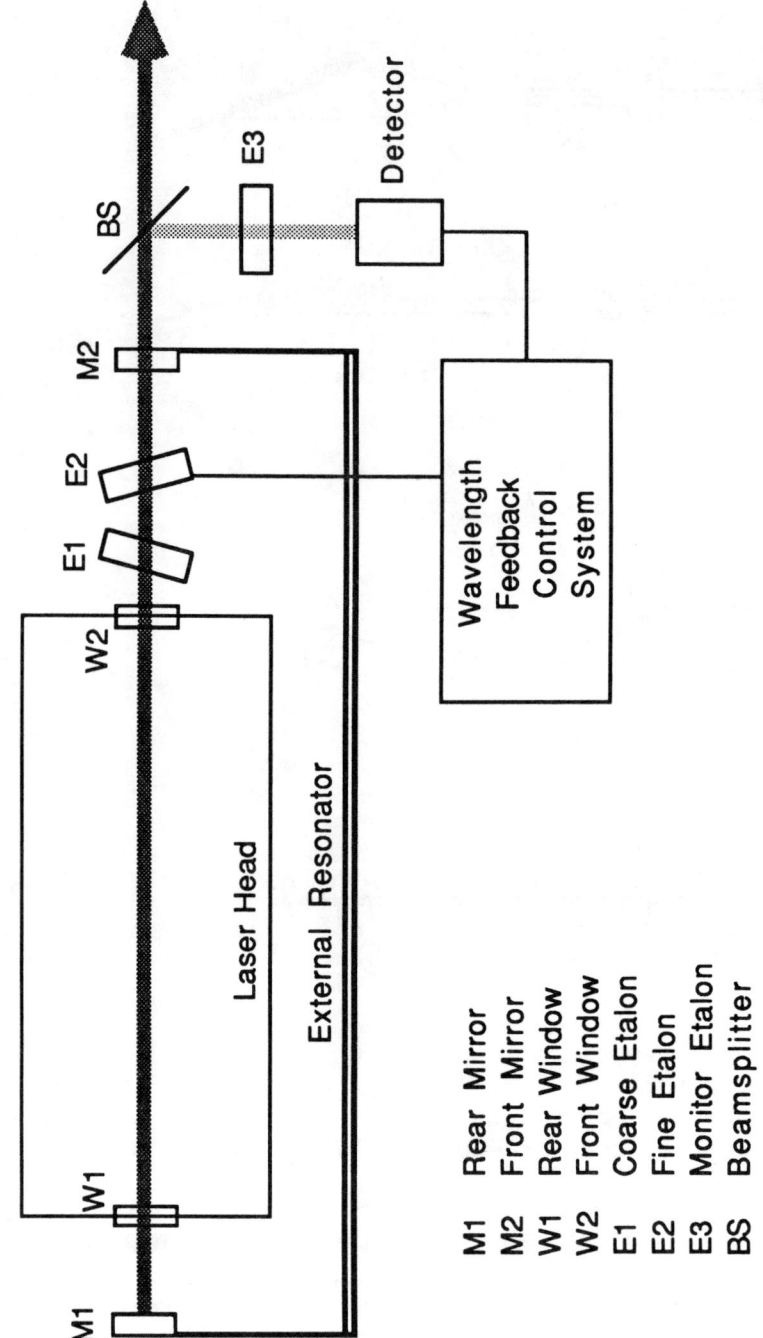

M1   Rear Mirror
M2   Front Mirror
W1   Rear Window
W2   Front Window
E1   Coarse Etalon
E2   Fine Etalon
E3   Monitor Etalon
BS   Beamsplitter

Fig. 5.9   Configuration for line-narrowing by intracavity etalons.

the center wavelength stabilized at a predetermined value. Commercial line-narrowed excimer lasers now offer a bandwidth of 0.003 nm with a center wavelength stability of 0.001 nm. Note that a bandwidth of 0.003 nm corresponds to a coherence length of $\sim$20 mm for a 248-nm KrF laser. Other performance specifications of these systems include an average power of 3 W, pulse repetition rate of 250 Hz, and pulse energy of >10 mJ. Looking at the developments in progress in various organizations, it appears likely that narrow-bandwidth excimer lasers with improved performance on all fronts will become commercially available in the near future.

An excimer laser, when line-narrowed by any of the methods shown in Fig. 5.8, not only becomes temporally more coherent, but its spatial coherence also increases to some extent due to the greater dependence of the mode structure of its output beam on the resonator optics. For example, it has been observed that exposures made with some spectrally narrowed lasers show evidence of speckle. In microlithography systems employing narrow-band excimer lasers, therefore, it is advantageous to incorporate certain modifications in the illumination system to eliminate the unwanted interference effects that arise from spatial coherence. These modifications, which are often necessary anyway for uniformization of the output intensity profile (see Sec. 5.3.4), usually consist of one or more of various beam-averaging devices such as stationary or rotating fly's-eye-lens integrators, diffusers, and scanning mirrors [36,55,68,74,75].

5.3.4. Beam Uniformization

In a practical lithography exposure system, the illumination is often required to be uniform within $\pm$2%. As can be readily observed from the intensity profiles shown in Fig. 5.3, a typical excimer laser beam, although closer to a flat-top than the outputs of most non-excimer lasers, is far from satisfactory in meeting the uniformity requirements of a production lithography system. We remark that the beam profiles of Fig. 5.3 are for a free-running laser, the intensity distributions for spectrally narrowed lasers being even less rectangular. Thus, it is necessary to design into the illuminator a beam uniformization system.

A number of different techniques for producing uniform intensity distribution have been developed for a variety of industrial optical systems, including conventional microlithography machines using Hg-Xe arc lamps as light sources. Many of these uniformization concepts are applicable in excimer laser illuminators. A most common beam-homogenization techniques employs a fly's-eye-lens integrator. An implementation of this approach in mercury-lamp illuminators, as found in Canon's conventional lithography tools, is shown in Fig. 5.10. The task of the two-dimensional lens array is to divide the incident beam into several components, expand each component so as to illuminate the whole mask, and superpose the components with each other to achieve the desired uniformity. Essentially the same concept has been used in the illumination systems of various excimer-laser steppers [55,74]. An example is presented in Fig. 5.11, which illustrates the use of a fly's-eye-lens integrator in the illumination system developed by Nikon for its excimer laser stepper [74]. In this particular case, the overall size of the lens array was 40x40 mm^2 and it consisted of 100 lenslets in a two-dimensional 10x10 array. Note that, as in the fiber optic beam transformation system illustrated in Fig. 4.14, the focal lengths of the individual lens elements in the present examples of Figs. 5.10 and 5.11 can be chosen to achieve the desired degree of partial coherence in the lithography system.

Another approach for beam uniformization consists in using a rectangular light pipe that has internally reflecting surfaces. In this method, the source beam is focused near the entrance of the pipe to create a cone of diverging rays, upon which multiple reflections from the pipe's internal surfaces transform the input beam into multiple virtual sources and produce the required superposition between different segments of the input beam. The spread of ray angles at the pipe entrance may alternatively be created by a diffuser. The light pipe

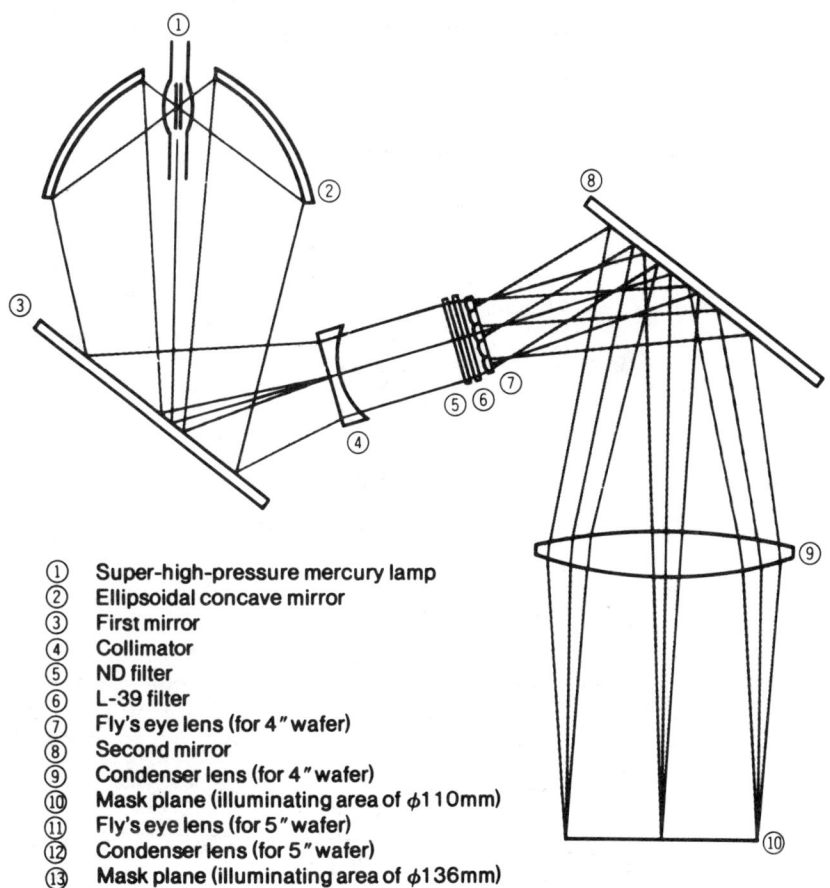

① Super-high-pressure mercury lamp
② Ellipsoidal concave mirror
③ First mirror
④ Collimator
⑤ ND filter
⑥ L-39 filter
⑦ Fly's eye lens (for 4″ wafer)
⑧ Second mirror
⑨ Condenser lens (for 4″ wafer)
⑩ Mask plane (illuminating area of φ110mm)
⑪ Fly's eye lens (for 5″ wafer)
⑫ Condenser lens (for 5″ wafer)
⑬ Mask plane (illuminating area of φ136mm)

Fig. 5.10   Illumination   system   with a   mercury   lamp
source used in   conventional exposure machines, showing
use of a   fly's-eye-lens   integrator for beam intensity
uniformization.   [Courtesy of Canon]

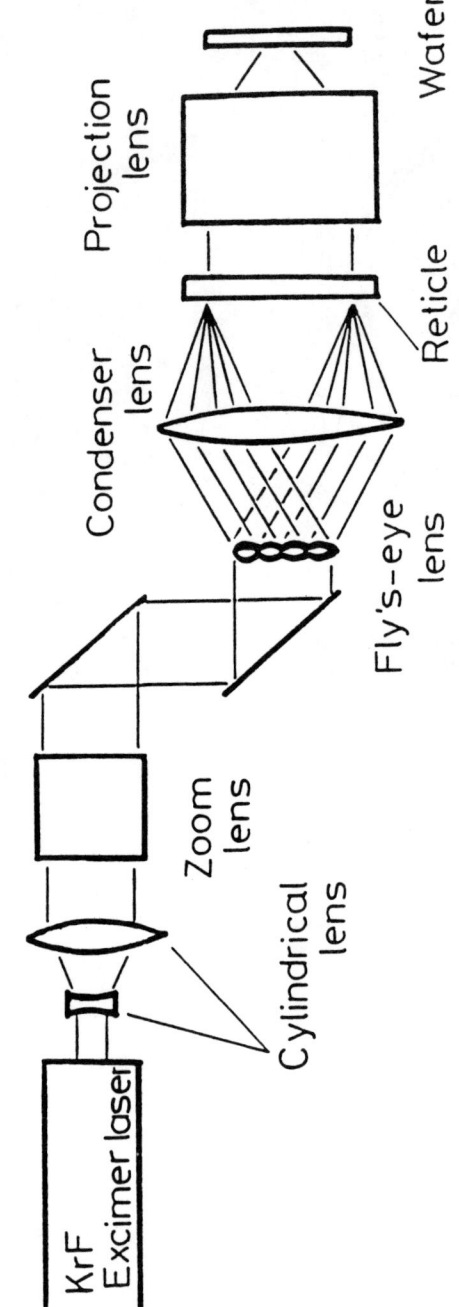

Fig. 5.11   Schematic illustration of the illumination system developed for an excimer laser stepper, showing use of a fly's-eye-lens integrator for intensity uniformization. [From Ref. 74]

may be either hollow or solid. A hollow light pipe is constructed with four front-surface mirror strips. A solid light pipe may simply be a rectangular quartz rod with optical-quality walls as well as input and exit faces. The uniformity achieved in the output beam is a function of the number of reflections within the pipe, which in turn is determined by the input cone angle and the dimensions of the pipe.

The light pipe uniformizer, like the fly's-eye-lens integrator, has been used in conventional lamp lithography systems. An example is shown in Fig. 5.12, which illustrates the illumination optics of several 1:1 Wynne-Dyson-type step-and-repeat tools developed by Ultratech. Figure 5.13 describes an application of a light pipe uniformizer in an excimer laser lithography system [75]. This illumination system was designed by Zeiss as a subsystem to go into ASM's excimer laser reduction stepper currently under development. In addition to a rectangular light pipe, it also contains a diffuser and a fly's-eye-lens integrator. In both Figs. 5.12 and 5.13, the light pipe is a rectangular, solid quartz rod.

Another intensity uniformization method, developed by Latta and Jain [13,111], uses a beam-folding wedged mirror tunnel. By folding the source beam about its axis several times, both long-range variations as well as short-range noise-like fluctuations in the given profile are averaged out. As the number of foldings increases, the degree of uniformization successively improves. The implementation of beam folding about its axis is accomplished using the basic reflection unit shown in Fig. 5.14(a). The collimated input beam is focused by a cylindrical lens into a line perpendicular to the plane of the paper and passing through point P. A mirror is positioned near P such that it is parallel to the above focal line as well as to the beam axis. The mirror position is designed to ensure that rays in the upper half-cone of the beam strike the mirror and are reflected, whereas rays in the lower half-cone travel past the mirror undisturbed. Thus, the incident cone angle, 2Θ, is cut in half and the input intensity

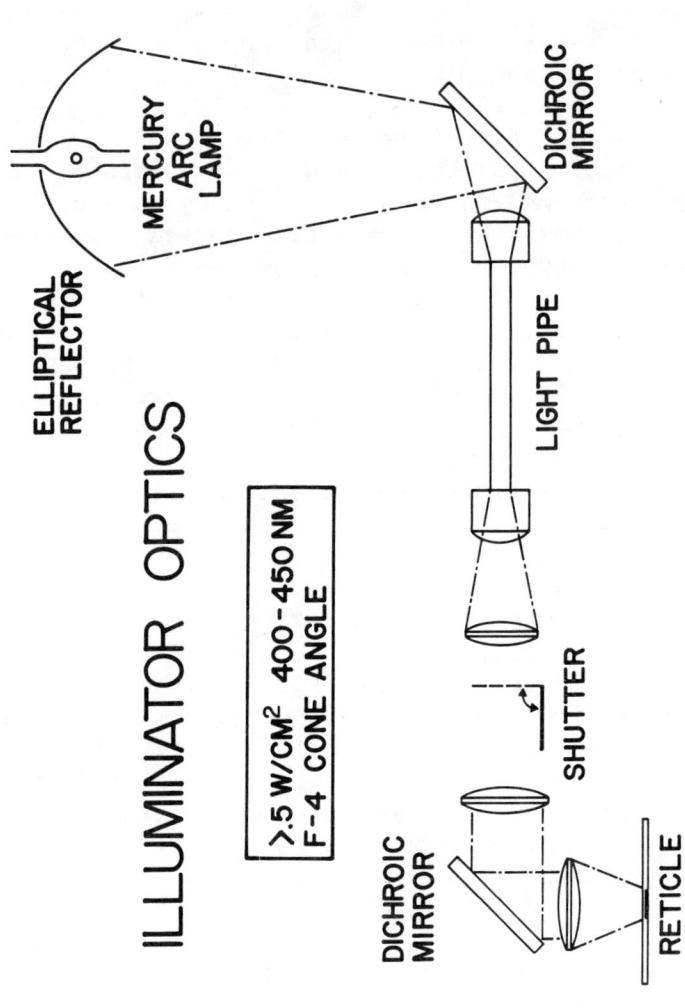

Fig. 5.12   Schematic of an illumination system showing use of a rectangular
light pipe for intensity uniformization. [Courtesy of Ultratech Stepper]

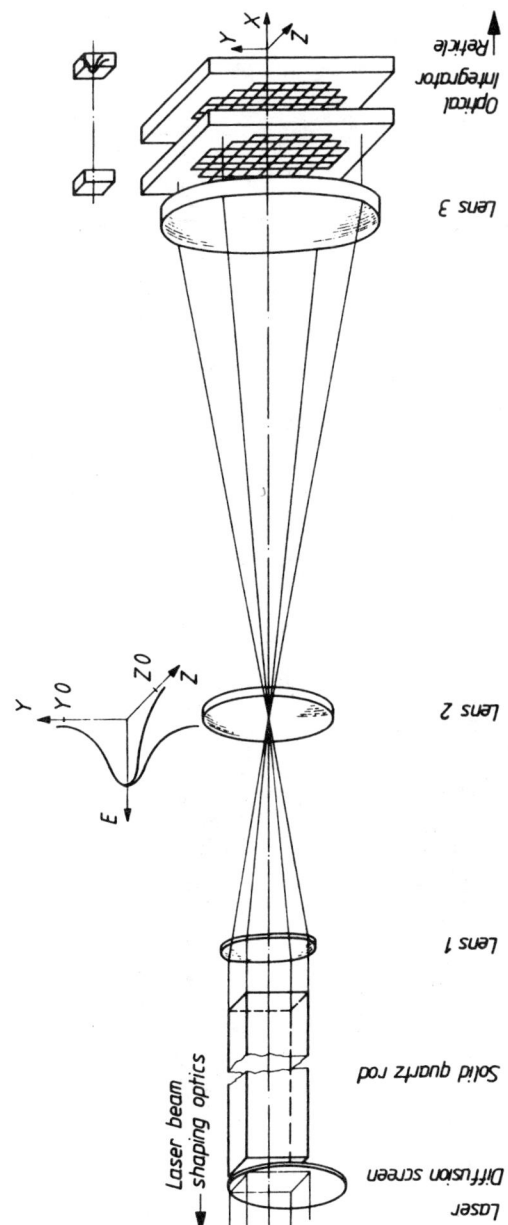

Fig. 5.13 Illumination system design employing a diffuser, a solid-quartz-rod light pipe and a fly's-eye-lens integrator for intensity uniformization. [From Ref. 75]

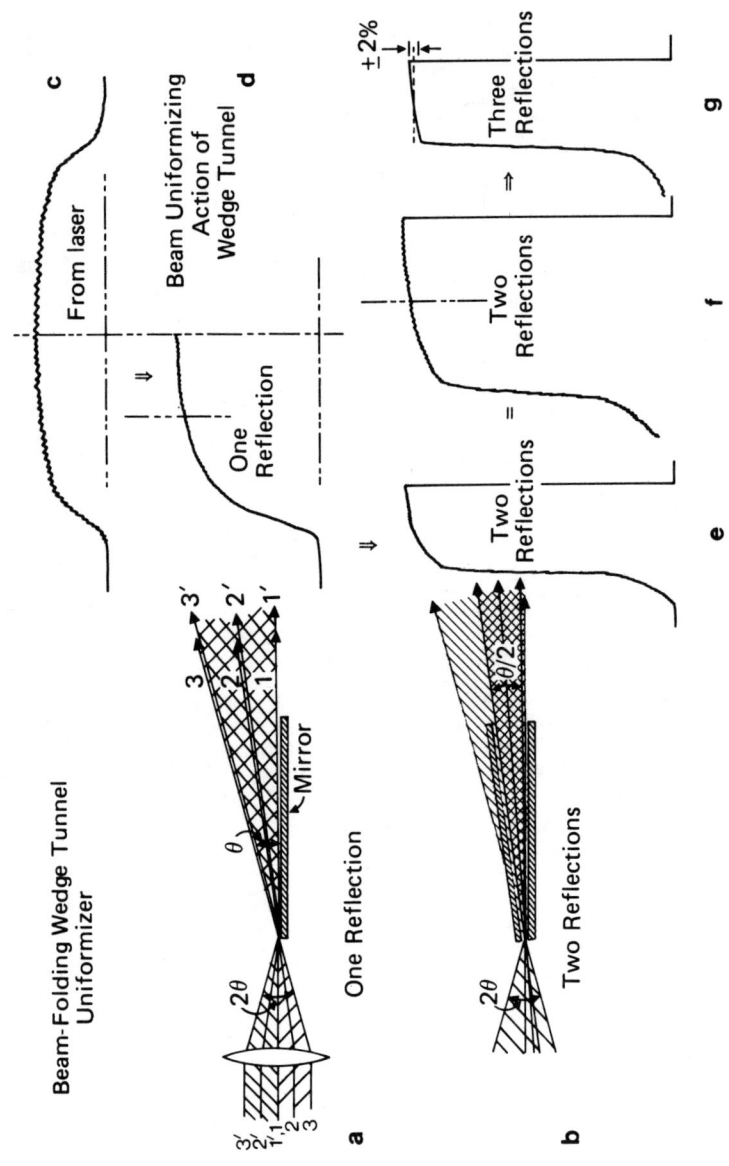

Fig. 5.14 A beam-folding wedge tunnel intensity uniformizer using (a) one mirror and (b) two mirrors. (c)-(g) Beam profiles after successive folding operations showing improvement in uniformity. [From Refs. 13,111]

distribution is folded onto itself. If desired, a
second cylindrical lens may then be used to recollimate
the output cone of rays.

    Multiple folds may be accomplished by forming a
wedge tunnel with two mirrors as shown in Fig. 5.14(b),
in which the upper mirror has been added to the basic
reflection unit of Fig 5.14(a). If the upper mirror is
inclined at an angle $\theta/2$ to the lower mirror, it will
bisect the output cone that would have been produced by
one mirror alone and thus create a second fold. Three
folds may be produced by decreasing the wedge angle to
$\theta/4$. In general, if one desires to fold the beam n
times, the angle between the mirrors should be $\theta/2^{n-1}$,
where 2θ is the incident cone angle. The improvement
produced in the beam profile upon successive foldings
is illustrated in Figs. 5.14(c)-(g). A two-mirror wedge
tunnel constructed in the above fashion and optimized
for use with an excimer laser beam profile resembling
that of Fig. 5.14(c) produced the output distribution
shown in Fig. 5.15, with an intensity variation of only
$\pm 1.5\%$ over a major portion of the beam containing 75%
of the total energy. Note that the configuration of
Fig. 5.14(b) provides intensity averaging in a single
dimension only. For uniformization in two orthogonal
dimensions, it is necessary to use two beam-folding
mirror tunnel units oriented with respect to each other
as illustrated in Fig. 5.16(a); alternatively, the two
tunnels may be combined into one unit as shown in Fig.
5.16(b).

    An advantage of the wedge-tunnel uniformization
technique is its high throughput, i.e., the flat-top
portion of the output profile contains most of the beam
energy. Another advantage is that the method does not
produce multiple virtual sources, nor does it destroy
the collimation properties of the input beam; as a
result, the output may be easily recollimated with a
simple lens. A possible limitation of the technique is
the creation of interference fringes when the input
beam is spatially too coherent; thus, the beam-folding
concept is especially suited for intensity-averaging
spatially incoherent laser beams.

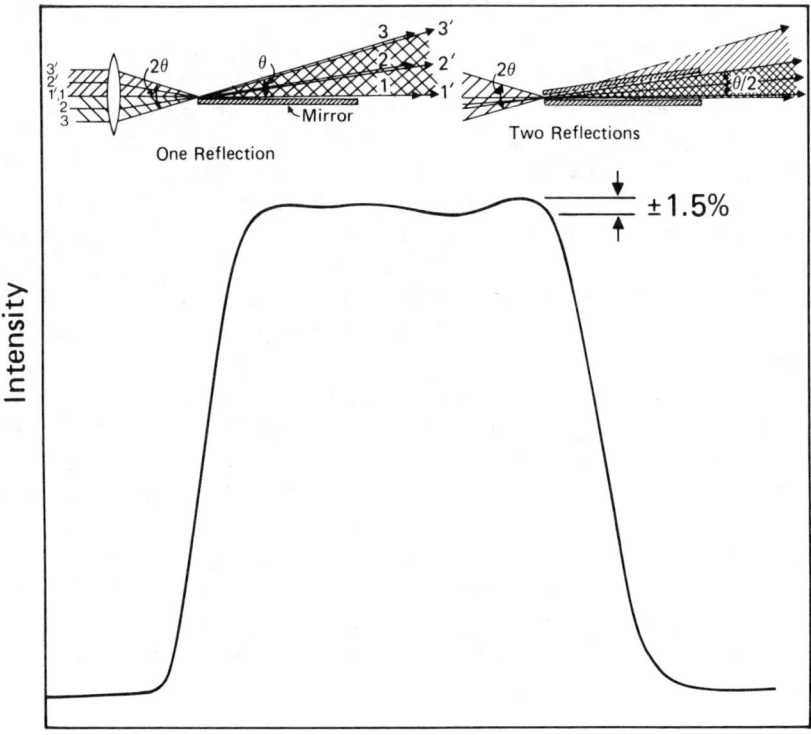

Fig. 5.15   Intensity uniformization by a beam-folding
wedge tunnel uniformizer. [From Ref. 13]

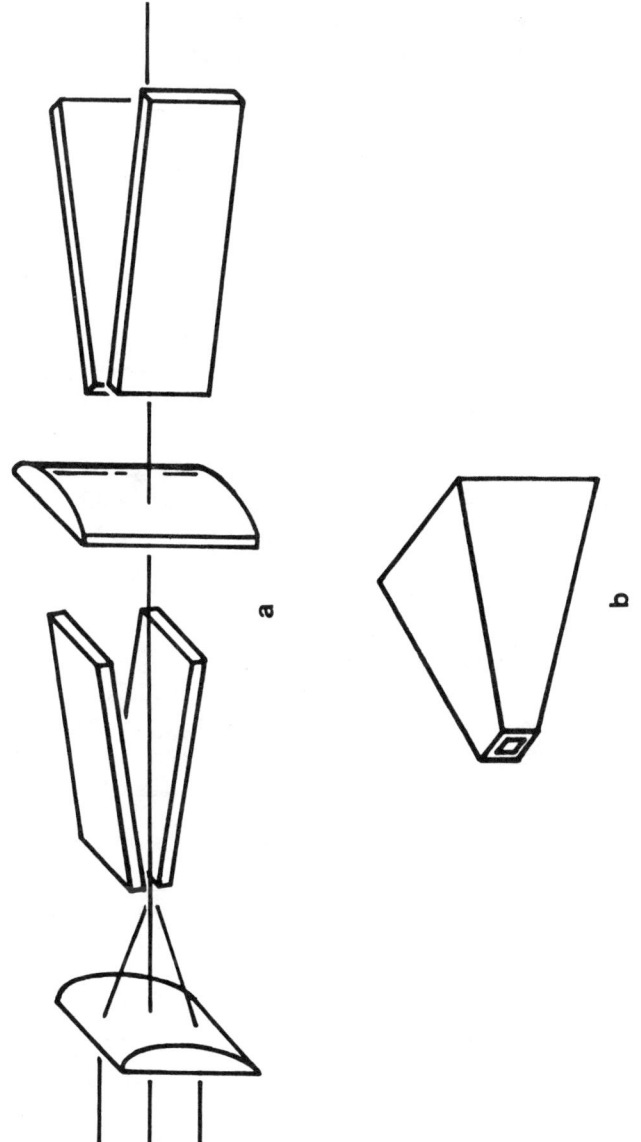

Fig. 5.16 Alternative configurations of a beam-folding wedge-tunnel beam uniformizer. [From Ref. 111]

A profile-reshaping technique using prisms [112] is shown in Fig. 5.17. Here, instead of folding the beam around its axis, it is split into four segments that are angularly displaced with respect to each other and then recombined to produce a more uniform profile. Note that the technique does not preserve the collimation properties of the input beam. It is also likely to produce, for spatially coherent input beams, interference fringes in the superposition region.

5.3.5. Maintenance and Reliability

For any manufacturing application of an excimer laser, and especially for an application as demanding as volume production of high-density microelectronic chips, the maintenance requirements of the laser, its reliability, and its long-term lifetime are important considerations. Unlike in most other lasers, the power output of excimer lasers declines rapidly with time due to the reactive nature of the halogen content of the discharge mixture. The performance degradation comes about as a result of the active gas mixture gradually deteriorating due to depletion of the halogen, as well as degradation of the laser optics from deposition of reaction compounds. As shown in Fig. 5.18, the output power of an excimer laser may typically drop by 50% in a day if no suitable measures are taken to compensate for, or arrest, these degradation mechanisms.

To remedy the above problems, several methods have been developed in the last few years. These include discharge-voltage ramping, periodic replenishment of the halogen, continuous gas mixture processing, and overall laser power feedback control. All of these features are now available in most commercial excimer lasers and readily provide stable performance over long periods of time. The high-voltage ramping technique consists in starting the laser operation at a discharge voltage lower than the maximum that the power supply can deliver and, then, as the output power begins to drop, ramping the voltage up to keep the output power constant. When the voltage can no longer be increased, the gas mixture is rejuvenated by partial replenishment

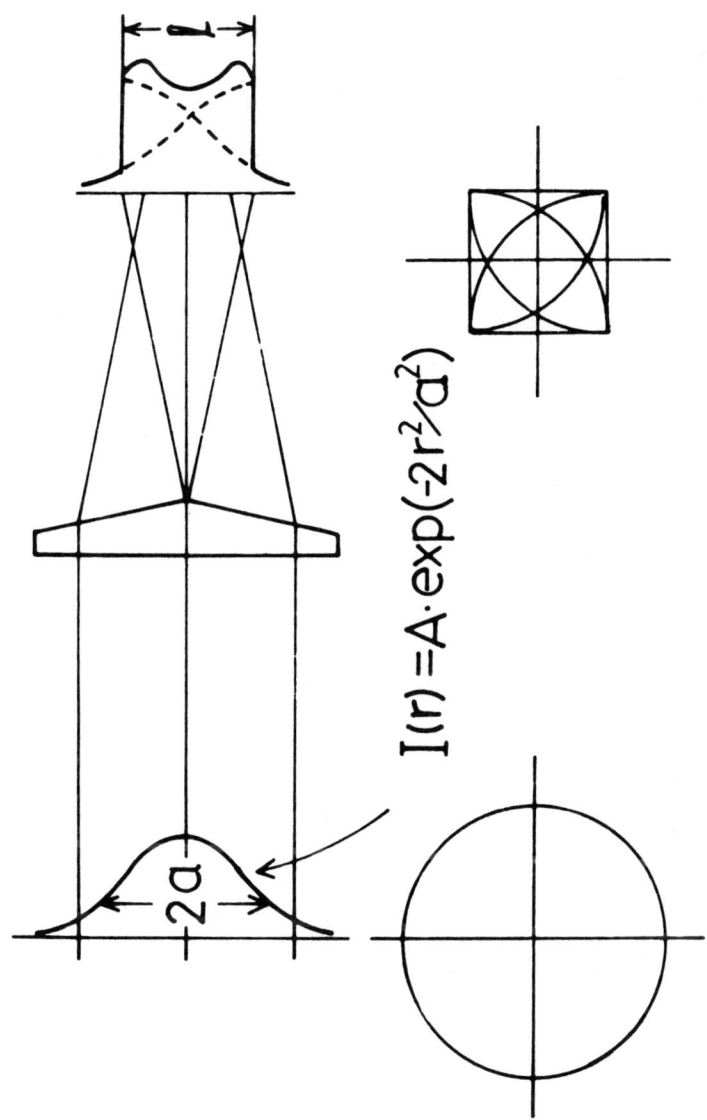

$I(r) = A \cdot \exp(-2r^2/d^2)$

Fig. 5.17 Use of prisms for improving the intensity uniformity of a Gaussian beam. [From Ref. 112]

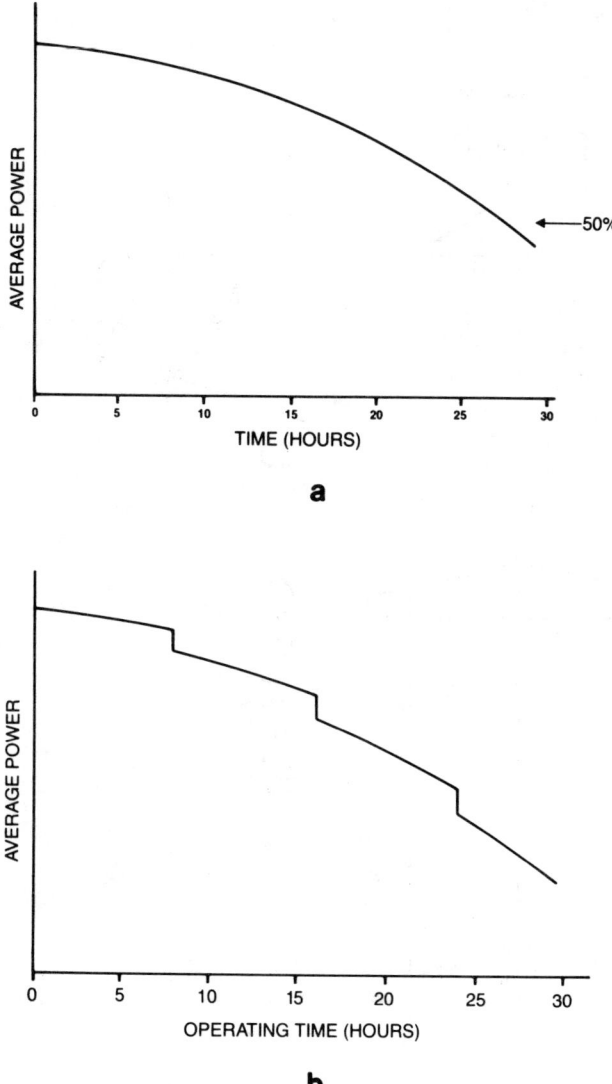

Fig. 5.18   Degradation with time in output power of an
excimer laser without power stabilization if the laser
is operated (a) continuously and (b) discontinuously.
[Courtesy of Questek]

of the halogen, as shown in Fig. 5.19. In addition, the entire gas mixture may be continuously or periodically purified and recycled by a gas processor containing a particulate filter and/or a liquid nitrogen cryogenic trap. The particulate filter eliminates solid impurity particles, usually down to 0.1-micron size or smaller, whereas the cryogenic trap helps remove most gaseous reaction products, which generally have a lower vapor pressure than the rare-gas and halogen constituents of the laser gas mixture. Further, some commercial excimer lasers also employ flushing of the window regions with an inert gas and miniature electrostatic precipitators to prevent contaminants from depositing on the windows.

Table 5.6 summarizes the performance status of currently available ArF, KrF, and XeCl excimer lasers. In the first row we show the number of pulses a laser delivers, without any stabilization feature, before its output power degrades by 50%. The second row gives the operating time in number of shots at constant power for a laser in which electronic feedback control of the power, partial gas replenishment, and continuous gas processing features have been added. In the third row, the above number of pulses has been translated into hours of operation at a repetition rate of 100 Hz with the laser running with a duty cycle of 50%. What is described in Table 5.6 is basically a single-gas-fill laser lifetime. For example, the 248-nm KrF laser may be run at constant power for 140 h with the stabilization features discussed above. After this period, the laser optics must be cleaned. This half-hour operation is required at most once in two weeks. The next level of maintenance involves servicing some electrical components in the laser, e.g., refurbishing or replacing the electrodes, and may be required once every few months. In summary, we remark that excimer laser performance and reliability have sufficiently matured for their incorporation into production-worthy manufacturing systems.

We conclude this chapter with a recapitulation of the relative significance of various excimer laser performance parameters as they relate to different

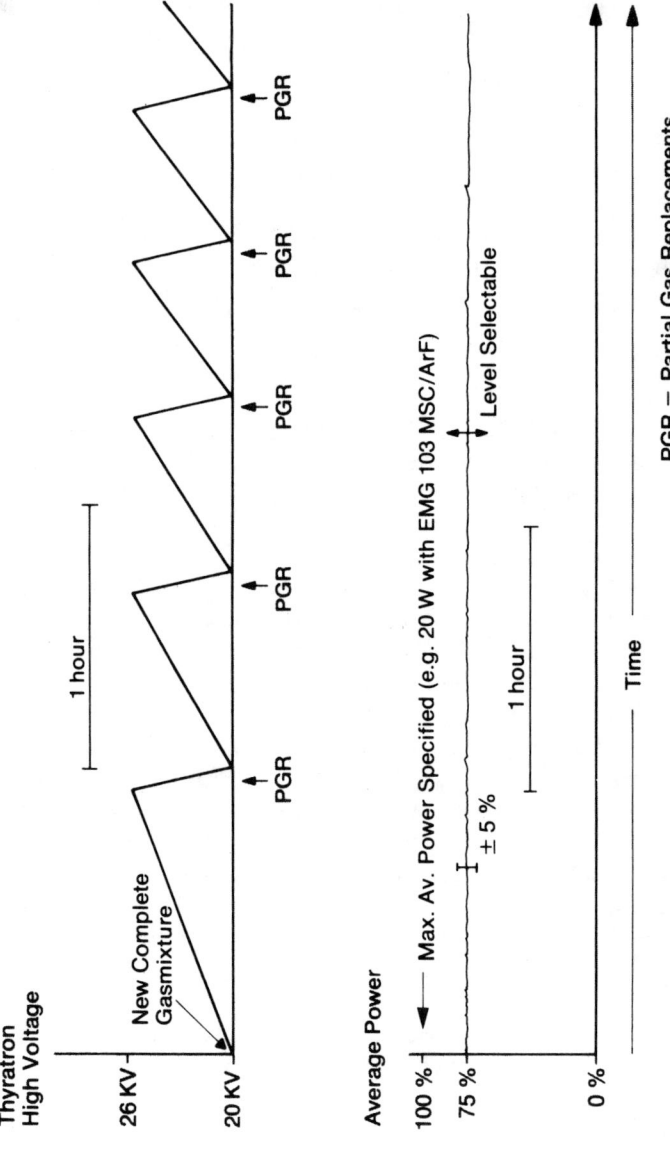

Fig. 5.19 Stabilization of the output power of an excimer laser with automatic feedback-controlled high-voltage ramping and periodic partial replenishment of the halogen. [Courtesy of Lambda Physik]

Table 5.6. Performance of excimer lasers:
output power stability and lifetime.

	ArF	KrF	XeCl
	193 nm	248 nm	308 nm
Single Gas-Fill Lifetime			
(No. of Shots to 50% Power without Stabilization)	6×10^5	2×10^6	2×10^7
Operating Time at Constant Power			
(with Electronic Laser Control, Gas Purifier and Partial Gas Replenishment)			
No. of Shots	2×10^7	2.5×10^7	4×10^7
Hours of Operation at 100 Hz (with 50% Duty Cycle)	110	140	220

lithographic systems. For scanning applications, as discussed in Sec. 4.2.1, the laser must be able to provide an output power of 5-10 W with a repetition rate of >200 Hz (the pulse energy being, therefore, a few tens of millijoules). Whenever possible, it is preferable to have, for a given average power, a higher repetition rate combined with a lower pulse energy. This is so because not only does a higher repetition rate lead to better exposure uniformity due to greater overlap between pulses, a lower pulse energy means lower peak power, which reduces the possibility of resist damage during exposure. Spectral narrowing of the laser is not a requirement for application in the mostly reflective 1:1 ring-field scanning systems since their bandwidths are far greater than the linewidth of a free-running excimer laser.

The laser requirements for step-and-repeat tools are quite different and vary from system to system, as discussed in Sec. 4.2.2. For conventional 5:1 or 10:1 refractive steppers, in which the projection lens consists of several elements, the most important laser parameter is the spectral bandwidth. If only one lens material (e.g., fused silica) is used, a line-narrowed laser with a bandwidth of <0.004 nm becomes necessary. In addition, tunability of the center wavelength within the gain spectrum of the lasing medium (which is ~1 nm wide for KrF, for example; see Fig. 5.5) is desirable. Such tunability is now available as part of the line-narrowing package in commercial units and facilitates optimum matching of the laser wavelength to the lens. On the other hand, in steppers using lens designs with two materials (e.g., quartz and LiF), and in systems based on the symmetrical 1:1 Wynne-Dyson design, one may use a conventional, non-spectrally narrowed excimer laser. As in the case of scanning tools, steppers will require lasers with repetition rates of a few hundred hertz, but greater pulse-to-pulse amplitude stability and spatial uniformity. The average power required from the laser for stepper applications is in the vicinity of 5 W. Finally, the pulse width of an excimer laser is not only hard to adjust, it is also not very relevant, except for the fact that it determines the peak power during exposure. If one had a choice of pulse widths, one would ideally want it as large as possible so that a given amount of energy could be delivered with the lowest possible peak power, thus reducing the susceptibility of the photoresist and the optical components in the exposure system to any possible damage by the laser pulses.

6. Resists for Excimer Laser Lithography

In addition to the projection system, the exposure wavelength, and the exposure source, an integral part of the total lithographic process is the resist medium and its processing, which help give physical shape to the microscopic circuit patterns on the wafer. In the progress in microlithography, as the minimum feature sizes in device geometries have shrunk with the use of shorter wavelength exposure tools, advances in resist technology including development of new materials have played an important role. The advent of excimer laser lithography and the wide recognition of its importance in continuing the dominance of optical lithography have given a special impetus to the study and development of resist materials and processes for use with excimer laser illumination systems. In this chapter we discuss various aspects of the technology of photoresists for excimer laser lithography. We present a discussion of several resist materials, their imaging properties at different wavelengths, dose sensitivities, and requirements for any special processing conditions. We also describe the reciprocity behavior of resists under excimer laser illumination and discuss the related photochemical aspects of resist exposure to high-peak-power laser pulses. The chapter places special emphasis on resists for use in the deep UV (DUV) with 248-nm KrF laser illumination systems.

6.1. RESISTS FOR EXPOSURE AT 308 NM

The most common positive-acting photoresists used commercially at conventional UV wavelengths, such as AZ 1350J and HPR 204, consist of two components: a matrix resin soluble in an aqueous base solution and a photo-active compound that acts as a dissolution inhibitor for the resin. The most widely used resin is a cresol-formaldehyde novolak resin, shown in Fig. 6.1(a), and the photoactive compound a substituted diazonaphtho-quinone, shown in Fig. 6.1(b). Upon absorption of UV light, the photoactive compound undergoes a structural

132

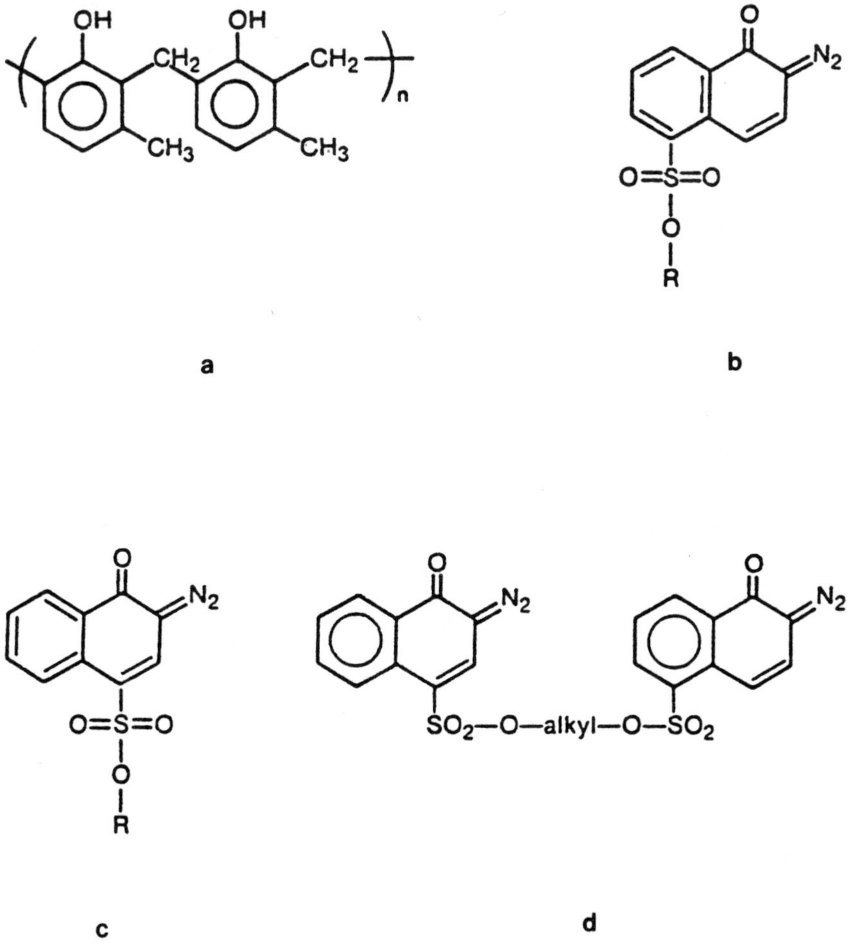

Fig. 6.1 Ingredients of mid-UV resists: (a) cresol-formaldehyde novolak resin; (b) 1-oxo-2-diazonaphtho-quinone-5-arylsulfonate sensitizer used in AZ 1350J; (c) 1-oxo-2-diazonaphthoquinone-4-arylsulfonate sensi-tizer used in AZ 2400; and (d) a mixed 4,5-disulfonate sensitizer used in an IBM mid-UV resist ER1.

transformation, which is followed by reaction with
water to form a base-soluble indenecarboxylic acid
[113]. The latter permits dissolution of the novolak
resin in alkaline developers and, thus, renders the
exposed resist areas more soluble than the unexposed,
making it possible to generate positive-tone images.

In mid-UV applications, the commercially available
conventional UV resist formulations based on the resin
and the photoactive compounds shown in Figs. 6.1(a) and
(b) do not perform as well as they do at conventional
UV wavelengths. The degradation in performance results
from the low sensitivity of the 1-oxo-2-diazonaphtho-
quinone-5-arylsulfonate photoactive compound, combined
with a significant unbleachable absorbance of the resin
in the mid-UV. Highly successful modifications in the
above formulation have been made to produce photoresist
systems with excellent mid-UV lithographic performance
by using more transparent novolak resins together with
more sensitive photoactive compounds, such as 1-oxo-2-
diazonaphthoquinone-4-arylsulfonate [Fig. 6.1(c)], as
in AZ 2400 [114], and a mixed 4,5-disulfonate of an
aliphatic diol [Fig. 6.1(d)], as in ER1, an IBM resist
[115,116].

Photoresists for excimer laser lithography at the
mid-UV wavelength of 308 nm emitted by the XeCl laser
have been successfully demonstrated in both contact and
projection printing modes [1-3,5,8,11,20,25,29,34]. As
previously described in detail in Secs. 4.1 and 4.2.1,
excimer laser exposures at 308 nm are characterized by
excellent imaging results, as well as by exposure dose
sensitivities similar to those normally observed in
microlithography with mercury lamp illumination. These
results have been demonstrated extensively in Shipley/
AZ 2400 and other common mid-UV resist systems
frequently used in production wafer lithography with
conventional Hg-lamp tools. Examples of images obtained
by contact printing at 308 nm with the XeCl laser in AZ
2400 resist are shown in Fig. 4.1 [1]. The scanning-
electron micrographs in Figs. 4.10 and 4.11 [29] show
mid-UV excimer laser projection images patterned on,
respectively, Perkin-Elmer Models 500 and 111 1:1 full-

wafer scanning machines in an IBM resist ER1, which is a formulation of the diazonaphthoquinone-novolak type, similar to AZ 2400. Note in all cases - Figs. 4.1, 4.10 and 4.11 - the excellent images with near-vertical wall profiles. In addition, in all mid-UV resists, it was also determined that little or no reciprocity failure takes place in 308-nm excimer laser exposures at doses of lithographic interest. Thus, it may be asserted that from the point of view of lithographic imaging, the behavior of the above mid-UV resists is essentially the same under excimer laser and Hg-lamp illumination. The reciprocity behavior of the resist process, including its lithographic, optical, and photochemical aspects, is discussed at length in Sec. 6.4.

6.2. RESISTS FOR EXPOSURE AT 248 NM

A large number of chemical formulations have been investigated as possible photoresist candidates for illumination with deep UV wavelengths in the $\sim$250-nm region, particularly with the 248-nm KrF laser. The comprehensive picture that emerges from these studies is that the overall photoresist situation for 248-nm exposure is significantly different from that for 308-nm exposure. The primary differences between deep UV and mid-UV excimer laser exposures of resists arise from the differences in their optical and photochemical behaviors in the two spectral regions. The performance of most of the commercially available resist materials is considered unacceptable at deep UV wavelengths due to certain fundamental shortcomings exhibited by them: resists that have good plasma-etching resistance show excessive unbleachable absorbance, whereas those with better transmission characteristics have inadequate etching resistance. In a photoresist with high deep UV absorbance that does not bleach, as in the well-known diazonaphthoquinone-novolak systems, the light exposure intensity received near the bottom of the resist layer is much smaller than that near the top; this invariably produces overcut image profiles with sloping walls. Consequently, although deep UV projection lithography systems can readily deliver image resolution in the

vicinity of 0.5 micron, the maximum resist thickness
that one may use with conventional resists for deep UV
exposure must be limited to below 0.5 micron. Such a
resist thickness value is too low from two points of
view: first, it provides inadequate coverage of the
underlying topography on the wafer, and second, its
etching resistance is poor. In the sections below we
describe various categories of deep UV resists and the
approaches investigated to overcome the limitations
mentioned above.

6.2.1. Dissolution Inhibition Resists

Exposures with the 248-nm KrF excimer laser in a
variety of photoresists have been reported in contact
printing mode [1-3,5,8,15,19,20,27,39,48], 1:1 full-
wafer scanning projection printing mode, and in step-
and-repeat projection printing mode [26,37,45-47,49,55-
59,63, 65,68,70,72]. Some of these have been previously
described in Secs. 4.1, 4.2.1, and 4.2.2. Examples of
images contact-printed with a KrF laser in Shipley/AZ
2400, the most widely studied dissolution-inhibition
deep UV resist system, are shown in Fig. 4.2 [1]. These
images were produced in 1-micron-thick resist with an
exposure dose of 125 mJ/cm^2. The expected sloping wall
profiles are clearly seen in these scanning electron
micrographs. Patterns printed by projection lithography
systems exhibit even greater slopes. Some examples
of projection images obtained in AZ 2400 are given in
Figs. 4.12, 4.17 [7] and 4.25 [57]. As mentioned in the
preceding paragraph, the wall slopes can be improved by
using thinner resist layers; further, overexposure also
helps in making the profiles more vertical. This is
illustrated in Fig. 4.3, which shows contact-printing
results in 0.4-micron-thick Shipley 2400-17 resist in
which an exposure dose of 230 mJ/cm^2 was used [27].

Many other novolak-based and other dissolution-
inhibition-type resist systems have been investigated
for 248-nm excimer laser lithography. These include AZ
4050 [27] and 5214 [49], McDermid PR-1024MB [49], Hunt
HPR 1182 [57], and Hitachi RD-2000N [49,57] and 5000P
[27]. AZ 4050 is designed primarily for use at near UV

wavelengths and has higher absorbance than Shipley/AZ 2400 at 248 nm; it therefore exhibits image profiles of poorer slope characteristics. 0.5-micron line-space pairs produced in AZ 4050 by contact printing with a KrF laser are shown in Fig. 6.2(a) [27]. AZ 5214, Hunt HPR 1182, and McDermid PR-1024MB, although designed for different process conditions, have deep UV absorption properties similar to AZ 4050 and thus produce similar images in single-layer lithography experiments [49,57]. Hitachi RD-2000N, which consists of an azide sensitizer and a phenolic resin, also absorbs heavily at 248 nm, but since it is negative-acting, it produces undercut profiles [49, 57]. The formulation of the Hitachi 5000P resist is somewhat different from other novolak-based resists in that its base resin is pure para-cresol novolak, whereas the other novolak-based resists use various mixtures of ortho-, meta-, and para-cresols. The use of pure para-cresol novolak produces significantly higher transmission at 248 nm; however, this also leads to exposure energy dose requirements as high as 2 J/cm^2. In addition, the 5000P suffers from a high development rate in the unexposed regions, leading to poor image profiles. An example of contact-printed images in this resist is shown in Fig. 6.2(b).

6.2.2. Organosilicon Resists

The fundamental problem encountered in KrF laser exposure of the otherwise highly desirable positive photoresists with novolak resin and diazonaphthoquinone sensitizer is that, as discussed above, the deep UV absorbance of both of its components is high, typically >1 micron^{-1}, and does not bleach on exposure. Thus, a logical remedy to the above hurdle would be to seek a chemical formulation with greater transmission than novolak and diazonaphthoquinone in the deep UV. Some success has been achieved in this direction by Orvek et al., who have investigated a positive-acting organo-silicon resist [48]. The base resin, PVPTMS, in this system is O-trimethylsilyl poly(vinylphenol), which is used with an organic sensitizer that acts through photogeneration of halogen acids that catalytically convert the PVPTMS in the exposed resist areas into a

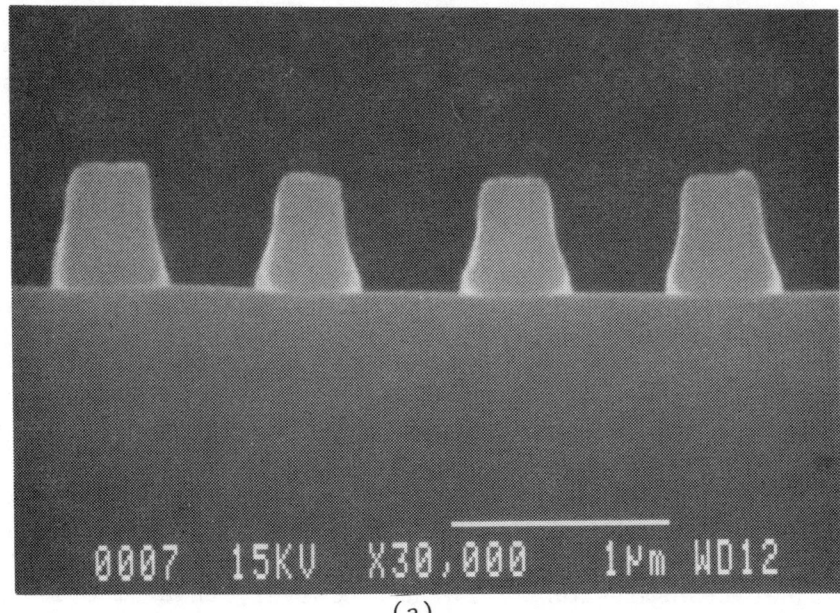

(a)

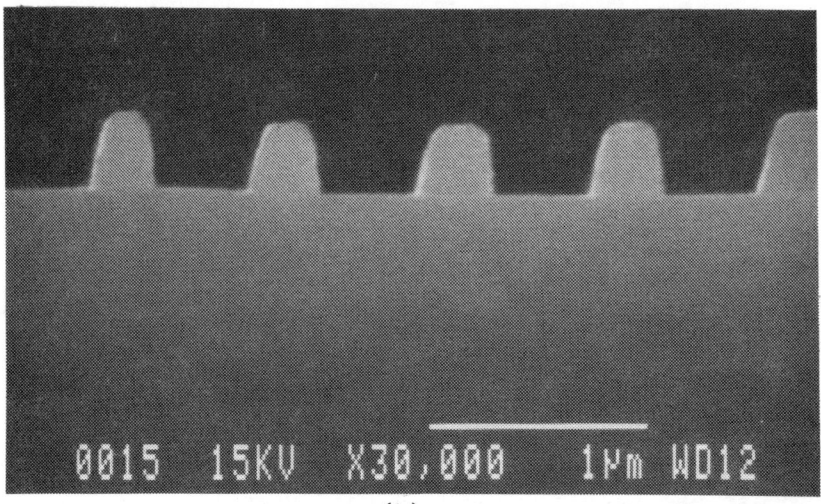

(b)

Fig. 6.2 Scanning electron micrographs of patterns contact-printed with 248-nm excimer laser exposures: (a) 1.0-micron-pitch features in AZ 4050 resist and (b) 0.8-micron-pitch features in Hitachi 5000P resist. [From Ref. 27]

material with a significantly increased concentration
of hydroxyl groups, making it more soluble in aqueous
developers, as illustrated in Fig. 6.3. A comparison of
the 248-nm transmission characteristics of the PVPTMS
resist with those of two novolak-based Shipley resists
- S2400, already discussed above, and S1400, a common
g-line resist - is shown in Fig. 6.4(b). Note the
markedly higher transmission of PVPTMS, which, as may
be expected, should result in images with more vertical
wall profiles than AZ 2400. An example is given in
Fig. 6.4(a), which illustrates contact-printed images
obtained in 1-micron-thick PVPTMS with 248-nm KrF laser
exposure. Although these imaging characteristics appear
satisfactory, this resist system is not close to being
usable in a production environment, particularly due to
problems associated with deposition of residue during
development and postexposure chemical activity of the
photogenerated acids. PVPTMS has also been used in a
bilayer patterning scheme, which is discussed in Sec.
6.2.5.2.

6.2.3. Chain Scissioning Methacrylate Resists

Another class of positive-acting resist materials
that have been studied extensively in the deep UV is
the poly(methylmethacrylate) (PMMA) family of polymers.
Two examples of such compounds are shown in Fig. 6.5.
Upon absorption of deep UV and vacuum UV light, these
materials undergo a chain scissioning mechanism, which
leads to a reduction of their molecular weight and, as
a result, to increased solubility in the exposed resist
areas. Thus, photochemically, the methacrylate polymers
function differently from the dissolution-inhibition-
type resists such as the diazonaphthoquinone-novolak
systems. Polymers in this class are characterized by
good transmission in the deep UV, but are unacceptable
for most production lithography applications because
their plasma-etching resistance is insufficient. The
optical transmission properties of PMMA and several
modified methacrylate copolymers have been measured by
Wolf et al. [39] and are shown in Fig. 6.6(a). These
should be compared with the transmission curves for
various conventional novolak-based resists, shown in

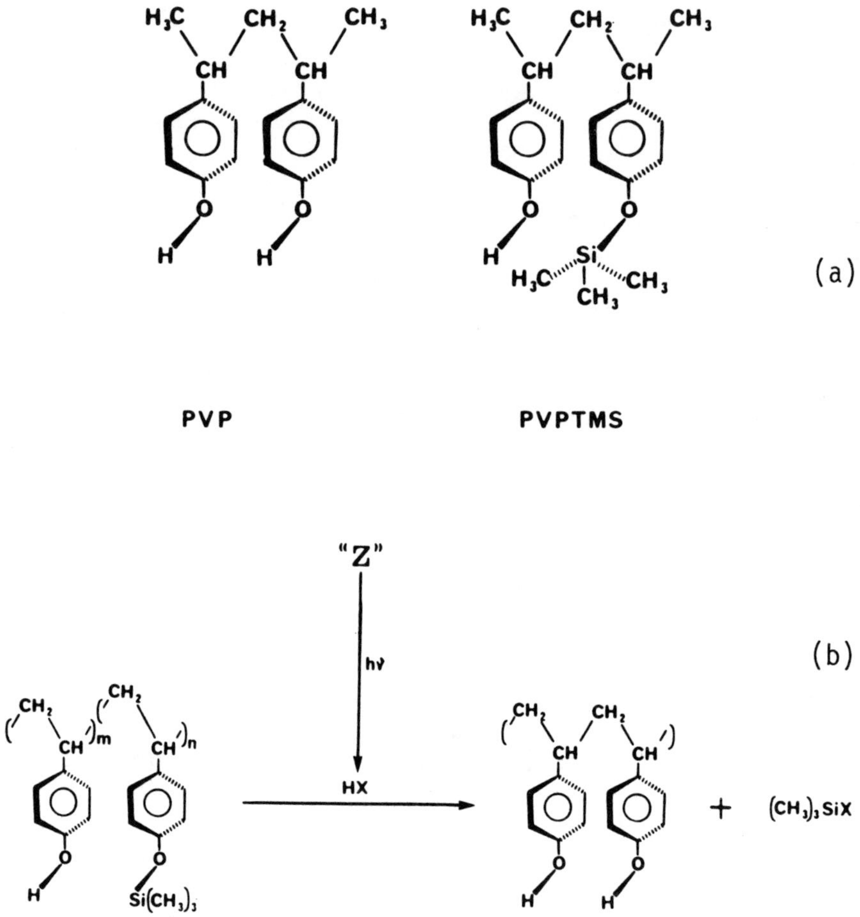

Fig. 6.3   (a) Structure of poly(vinylphenol) (PVP) and
the organosilicon derivative O-trimethylsilyl poly-
(vinylphenol) (PVPTMS);   (b) catalytic decomposition of
PVPTMS in the presence of photogenerated acid. [From
Ref. 48]

(a)

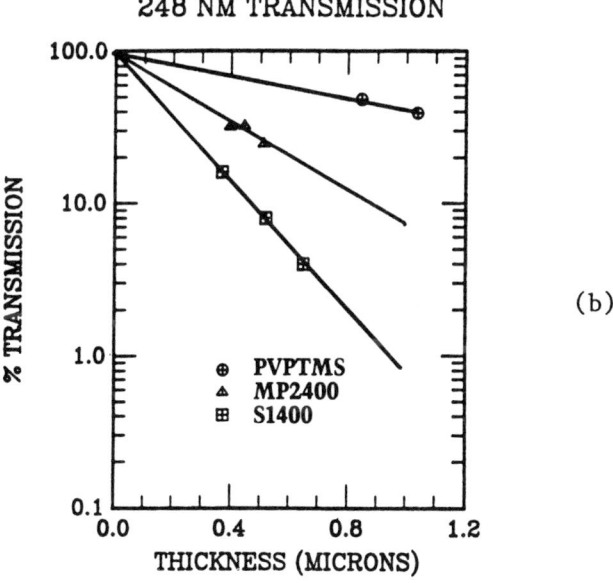

(b)

Fig. 6.4   (a) 0.5-micron images   contact printed   with
248-nm   KrF   excimer laser exposure in 1.0-micron-thick
PVPTMS resist;   (b) transmission   of PVPTMS and Shipley
S2400 and S1400 at 248 nm. [From Ref. 48]

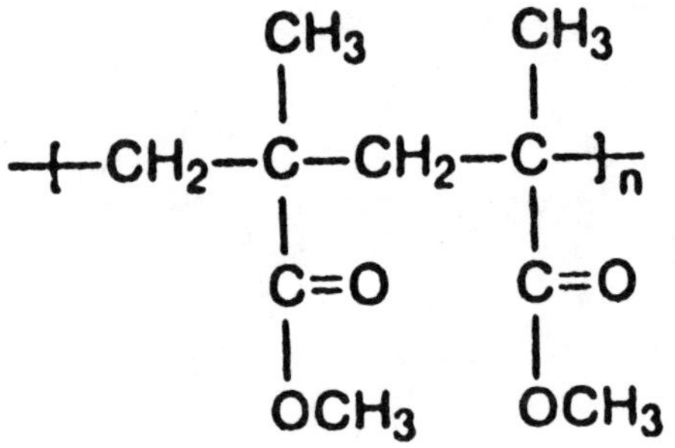

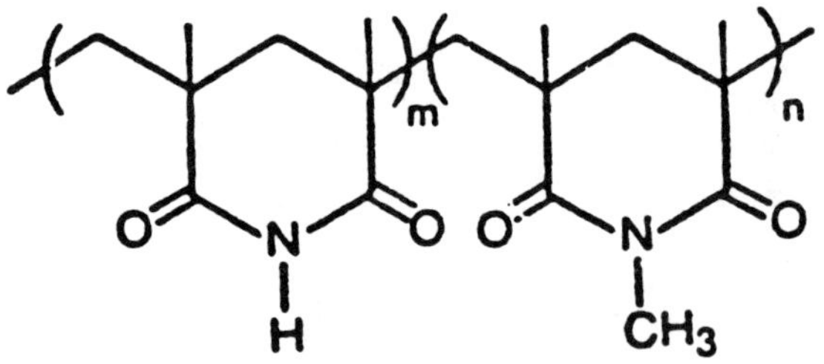

Fig. 6.5   Structure of (a) poly(methylmethacrylate)
(PMMA) and (b) poly(dimethylglutarimide) (PMGI).

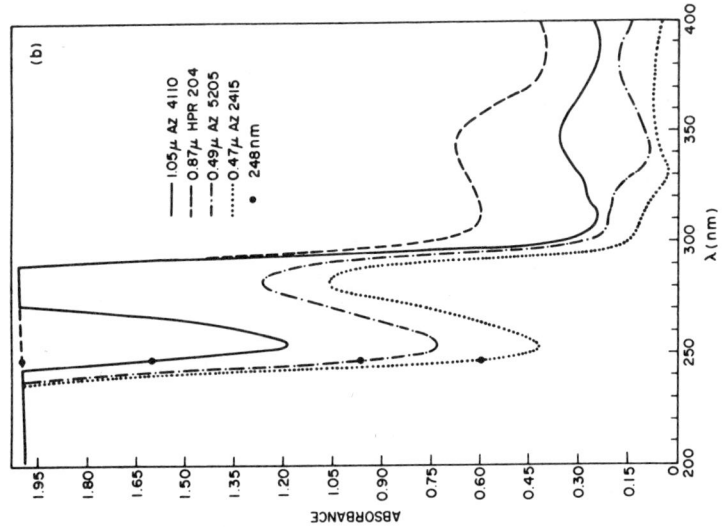

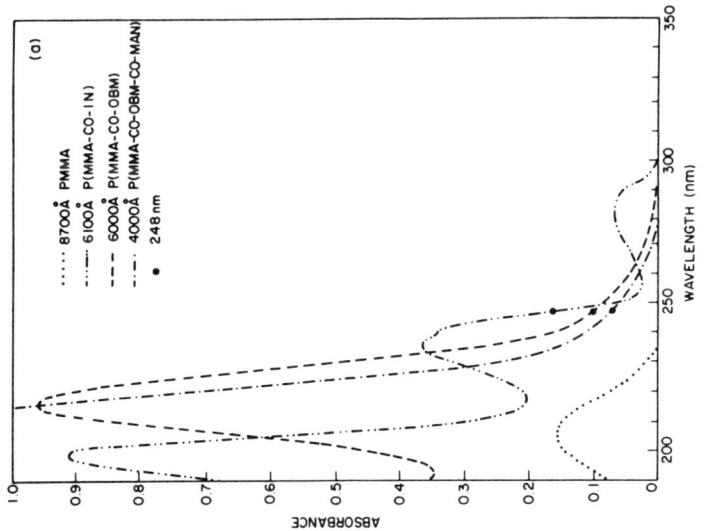

Fig. 6.6   UV absorbance spectra of several (a) methacrylate polymers   and (b) novolak-based resists. [From Ref. 39]

Fig. 6.6(b). As the data in these figures illustrate, the absorbance values for the methacrylate resists are an order of magnitude lower than for the novolak-based systems. As a consequence, although resists of the PMMA family require very high exposure doses, they produce images with excellent profiles. In projection printing experiments on a 0.37-NA 248-nm achromatic step-and-repeat system, Nakase et al. [47] have reported an exposure dose requirement of 32 J/cm^2 for PMMA and 10 J/cm^2 for poly(dimethylglutarimide) (PMGI). Kameyama and Ushida [46], using an identical exposure tool, have patterned 0.55-micron-wide lines and spaces in 1-micron thick PMMA and demonstrated that it is possible to obtain near-vertical profiles in this resist in deep UV projection printing; some examples of their results are shown in Fig. 4.26. We summarize the above discussion of novolak-based and other dissolution-inhibition-type resists and methacrylate-based chain-scissioning-type resists in Fig. 6.7 [49] with a comparison of the dose sensitivities and imaging characteristics alongside the optical transmission spectra for MP 2400, RD-2000N, PMMA, and PMGI.

6.2.4. Chemical Amplification Resists

Recent investigations in deep UV resist systems have directed attention toward chemical amplification as a means of achieving both high sensitivity and high resolution in single-layer patterning [63,65]. An IBM experimental deep UV resist, XPR, based on chemical amplification has been described recently by Woods et al. [63]. In this system, a photoactive precursor, on absorption of deep UV photons, undergoes decomposition and produces an acid. During a postexposure bake step, this photogenerated acid increases the solubility of a resin in aqueous alkaline developers, thus producing positive-tone images. In the presence of heat, the acid acts on the resist in a catalytic fashion. Thus, such a resist system is capable of exhibiting high sensitivities; exposure dose requirements as low as 10-20 mJ/cm^2 have been experimentally observed. The resin in XPR has an aromatic content, which gives the resist good plasma-etching resistance as well as good

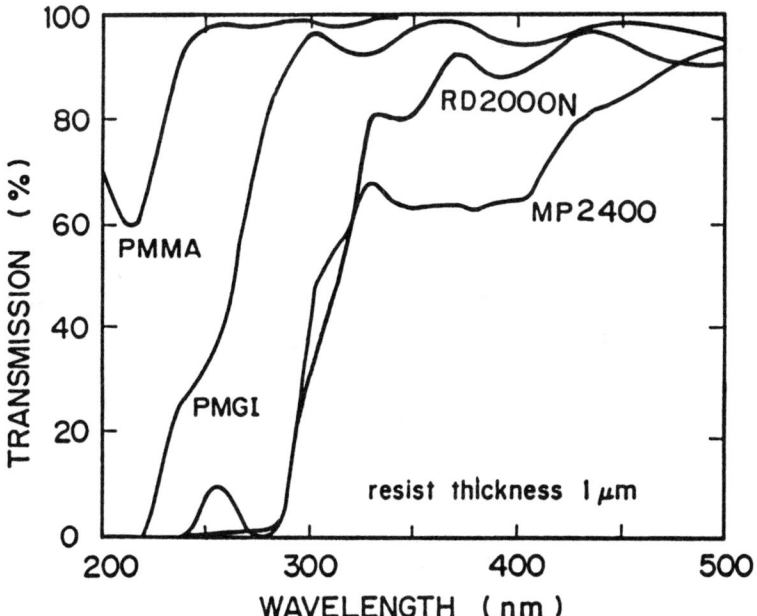

Fig. 6.7   Comparison of absorption, exposure dose sensitivity, and imaging characteristics of PMMA, PMGI, MP2400, and RD2000N resists.   [From Ref. 49]

thermal stability. In addition, the low (0.2 micron^{-1})
deep UV absorbance of the resin permits use of resist
layers of convenient thicknesses ($\sim$1 micron). Woods et
al. [63] have carried out an extensive lithographic
evaluation of XPR on a 0.35-NA monochromatic step-and-
repeat system. As an example, 0.5-micron-wide lines and
spaces with vertical wall profiles obtained in 0.85-
micron-thick resist are shown in Fig. 6.8.

Another chemical-amplification-based resist system
has been recently developed for commercial availability
in a joint effort between the Shipley Company and the
Rohm and Haas Company [65]. This negative-tone resist,
called XP-8843, consists of three components: a base
resin, poly(p-vinyl)phenol, chosen for its low deep UV
absorbance (0.17 micron^{-1}), good resistance in plasma
etching, and solubility in an aqueous base; a photosensitive acid generator; and a melamine crosslinking
agent. Absorption of deep UV light causes photolysis of
the photosensitive component to produce an acid; when
this is followed by a postexposure bake step at 110-
150 °C, a catalytic reaction takes place between the
resin and crosslinking agent resulting in a resist film
that is highly crosslinked in the exposed regions. The
greatly reduced solubility of the crosslinked resist
areas in an aqueous developer, thus, produces negative-
tone images. The above process flow is illustrated in
Fig. 6.9 [65]. A detailed lithographic investigation
of XP-8843, encompassing its optical, chemical, and
thermal aspects, has been reported by Thackeray et al.
[65]. The projection images in this study were printed
on a 0.35-NA monochromatic step-and-repeat tool. Figure
6.10 shows 0.5-micron line-space features obtained in
a 1-micron-thick resist film with a nominal dose of 16
mJ/cm^2. Note the near-vertical image profiles and the
large exposure dose latitude.

We remark that whereas resist systems based on
chemical amplification schemes appear attractive from
the points of view of dose sensitivity, image profiles,
and plasma-etch resistance, they also have the drawback
of requiring the extra postexposure baking step. The
latter may also be a potentially sensitive step, in

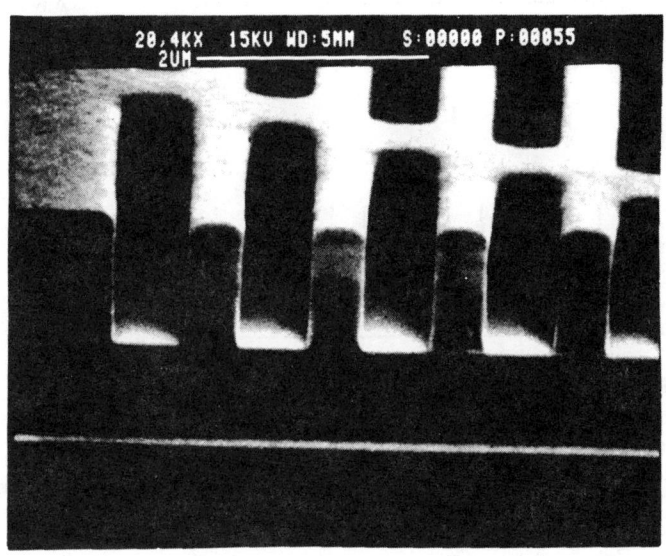

Fig. 6.8   Images with 0.5-micron line-space features
obtained in 0.85-micron-thick chemical-amplification
resist XPR with 248-nm excimer laser exposure on a
0.35-NA step-and-repeat system with a monochromatic
lens. [From Ref. 63]

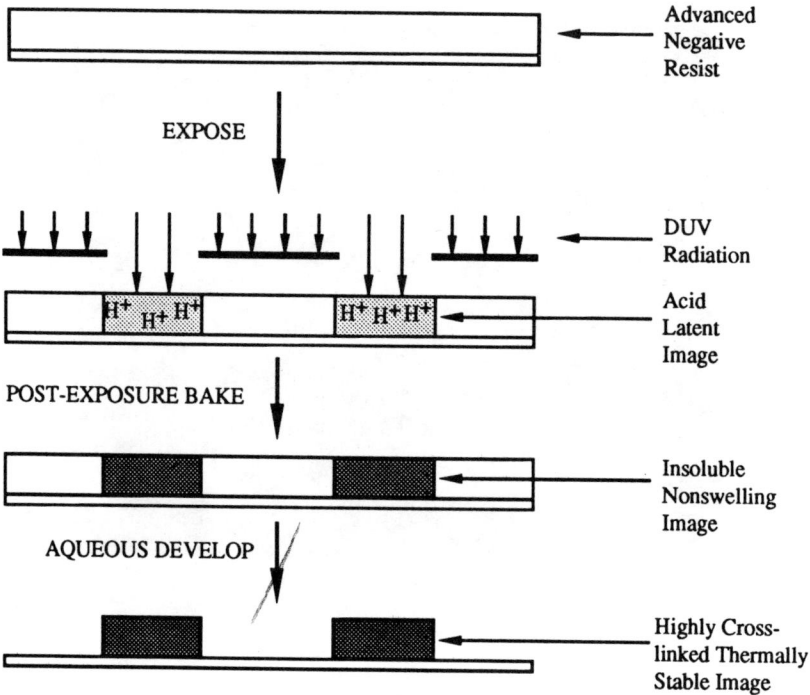

Fig. 6.9   Process flow for the Shipley XP-8843 deep UV
chemical-amplification resist. [From Ref. 65]

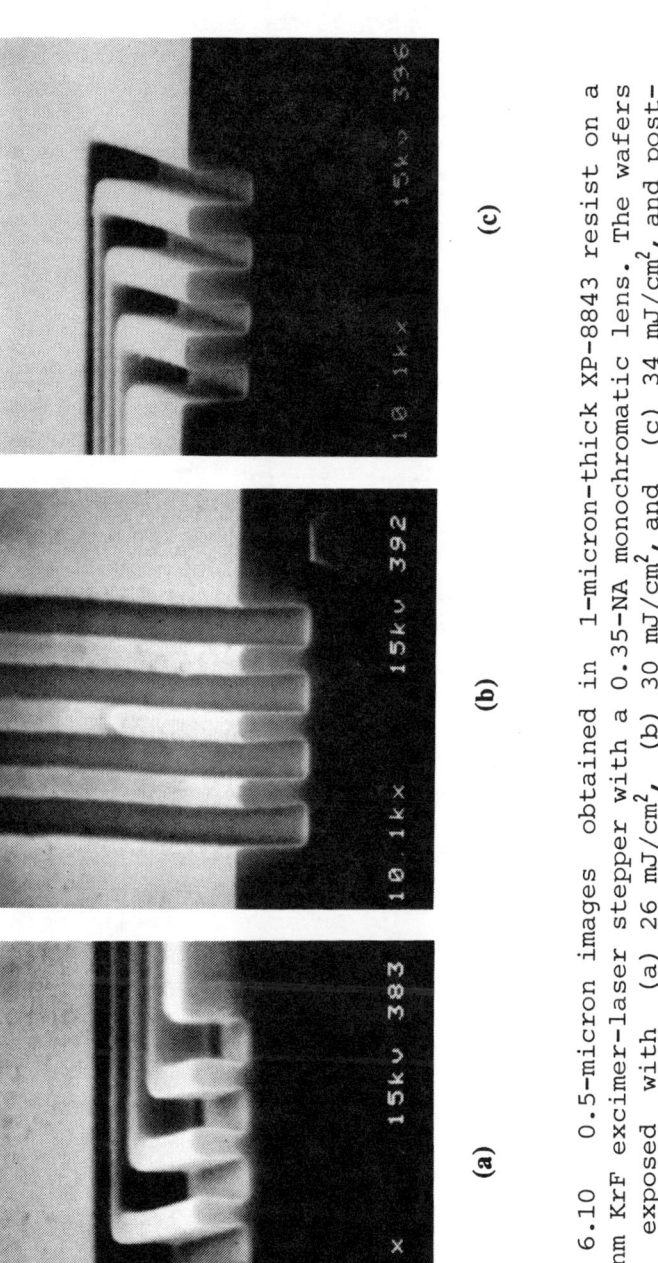

Fig. 6.10   0.5-micron images   obtained in   1-micron-thick XP-8843 resist on a
248-nm KrF excimer-laser stepper with a 0.35-NA monochromatic lens. The wafers
were exposed with (a) 26 mJ/cm^2, (b) 30 mJ/cm^2, and (c) 34 mJ/cm^2, and postexposure baked at 140°C for 1 min. [From Ref. 65]

that the dependence of the resist performance on it
is critical. It is expected that with more extensive
experimentation and a larger performance evaluation
data base obtained in production applications, a better
understanding of the key resist parameters will emerge.

6.2.5. Multilayer Resist Systems

In view of the various limitations of the deep
UV resists described above in single-layer production
lithography applications, there has been a sizeable and
ongoing effort in development of multilayer systems.
Although multilayer patterning clearly adds to wafer
fabrication costs due to numerous additional process
steps, its attractiveness arises not only from its
ability to produce high-resolution image profiles with
large aspect ratios, but also from the fact that it
potentially overcomes the fundamental depth-of-focus
limitation of optical lithography tools in single-layer
imaging. The importance of the depth-of-focus advantage
increases as the imaging wavelength decreases. Thus,
whereas in conventional UV and mid-UV lithography the
greater depth of focus offered by multilayer patterning
is a luxury one may decide not to have in view of the
added cost, in deep UV lithography many users view it
as an alternative that is very viable when the overall
economic considerations of the lithography process
steps in the wafer fabrication cycle are taken into
account. Further, in excimer laser lithography at the
shorter wavelengths of 193 nm and 157 nm, which are
beginning to be investigated for image resolution in
the 0.25-micron regime, it is likely that the use of
multilayer resist systems will become the preferred,
if not required, approach to pattern formation.

6.2.5.1. Trilevel Resist Structures

A very common multilayer resist system consists
of a three-layer structure: a top layer, <0.5-micron
thick, for imaging; a middle layer, $\sim$0.1-micron thick,
as an etch barrier; and a bottom layer, >1.0-micron
thick, into which the patterns are transferred by
dry etching. Using a 248-nm excimer laser stepper,

0.5-micron images have been obtained (Fig. 4.19) in a
trilayer structure consisting of a 0.5-micron-thick
imaging layer of Shipley MP 2400-17 on 1.8-micron-thick
hard-baked Hunt HPR 206 coated with 0.12 micron of SiO_2
[26]. As can be seen from Fig. 4.19, the well-known
sloped profiles that MP 2400-17 routinely produces in
deep UV exposures are absent in this trilevel imaging
scheme as a result of the use of a thin imaging layer,
and the fact that the final pattern is produced in HPR
206 by dry etching, which delivers vertical walls.

A trilayer resist system that uses an inorganic
imaging layer has been described by Polasko et al. [15,
19]. This system consists of a $Ag_2Se/GeSe_2$ imaging
resist on an AZ 1350J pattern-transfer layer. The $GeSe_2$
was deposited to a thickness of 0.18 micron on 1.5
microns of hard-baked AZ 1350J by thermal evaporation
from a bulk $GeSe_2$ source. A sensitizer film of Ag_2Se,
10-20 nm in thickness, was formed at the top surface
of the $GeSe_2$ by dipping the wafer in an aqueous $AgNO_3$
solution. Using a KrF excimer laser, deep UV exposures
of the sensitized layer were carried out at exposure
doses between 5 and 10 mJ/cm^2. Following exposures the
Ag_2Se was removed in dilute aqua regia and the $GeSe_2$
below was etched in a solution of $(CH_3)_2NH$, isopropanol
and water. The AZ 1350J may then be patterned by dry
etching. Figure 6.11 shows 0.5-micron lines and spaces
obtained in the $Ag_2Se/GeSe_2$ structure with an exposure
dose of 5.2 mJ/cm^2 at 248 nm [15]. It was found that
this inorganic resist system exhibits a large incident
dose nonlinearity and reciprocity failure; this aspect
is discussed in Sec. 6.4.

Recently, a trilayer resist system consisting of
AZ 5214, polysiloxane and OFPR-5000 has been studied by
Higashikawa et al. [49]. In this structure, following
imaging of the AZ 5214, the polysiloxane is reactive-
ion-etched in CF_4-H_2 and the pattern transfer into the
OFPR-5000 is done by reactive-ion etching in O_2.

6.2.5.2. Bilevel Resist Structures

In addition to the trilevel resists described

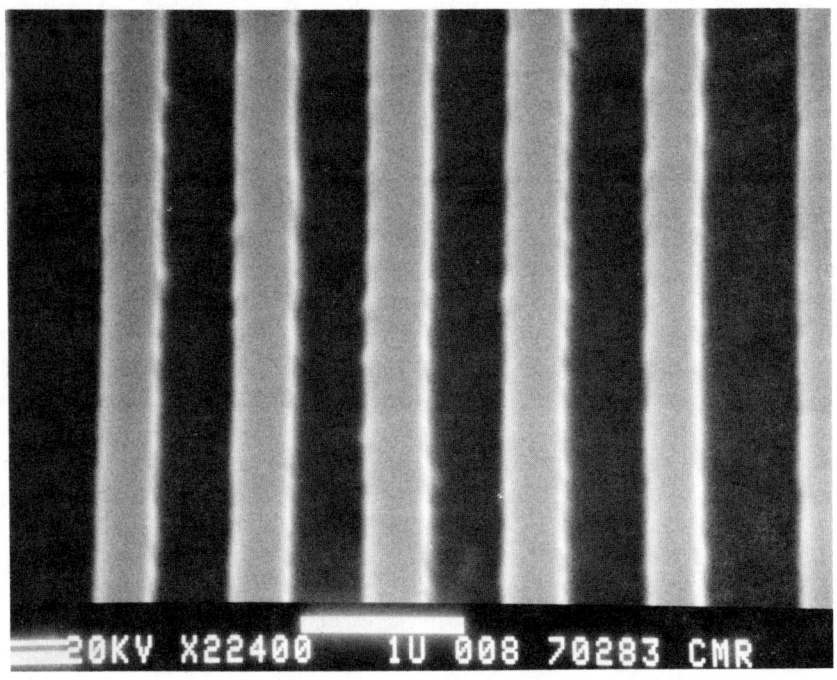

Fig. 6.11   0.5-micron images obtained in the inorganic resist $Ag_2Se/GeSe_2$ by excimer laser contact printing at 248 nm. [From Ref. 15, © 1984 IEEE]

above, several bilayer schemes have been reported in which the top imaging layer also serves as the etch barrier for pattern transfer into the bottom layer. Orvek et al. have used the organosilicon resist PVPTMS described in Sec. 6.2.2 in a bilayer configuration for lithography at 248 nm [48]. In this application, the PVPTMS imaging layer was 0.6-micron thick, coated over a 1.2-micron-thick planarizing layer of hard-baked Shipley S1800, a novolak-based photoresist, into which patterns were transferred by reactive-ion etching in O_2. Onishi et al. [117] have reported a bilayer resist structure consisting of a polysiloxane imaging layer on top of a hard-baked OFPR-500 pattern-transfer layer. Silicon-based positive- and negative-tone resists as the imaging layers in bilayer configurations have also been investigated by Kawai et al. [72]. An inorganic resist, peroxo-heteropolytungstic acid (HPA) on top of a polyimide resin, PIQ, has been studied by Ishikawa et al. [73]. In these experiments the 0.1-micron-thick HPA layer was imaged with a dose of 100 mJ/cm^2 by a KrF excimer laser, and the patterns were etched into 1.0-micron-thick PIQ by O_2 reactive-ion etching. Lines and spaces of 0.4-micron width were obtained with an aspect ratio of 2.5.

6.2.5.3. Portable Conformable Masking

Another non-single-layer patterning technique for deep UV lithography is portable conformable masking (PCM), pioneered by Lin [118]. This is also a double-layer method that employs a thin imaging layer on top of a thick pattern-transfer layer. But it differs from the concept behind the bilayer systems described in Sec. 6.2.5.2 above in that the pattern transfer into the thick bottom layer takes place by optical exposure rather than by dry etching. In the scheme described by Lin [118], the top layer is a conventional UV resist, such as AZ 1350J, which is highly absorbing in the deep UV. After imaging with conventional UV light, the top layer functions as a mask that is in perfect contact for deep UV flood exposure of the thick bottom layer. The latter is a suitable deep UV resist capable of producing vertical profiles in high-intensity flood

exposures, such as PMMA. Recently, a PCM resist system
in which both the top and bottom layers are exposed
with deep UV light has been reported [56]. In these
experiments, the structure consisted of a 0.5-micron-
thick top layer of Hitachi RD-2000N, and a 1.0-micron-
thick bottom layer of PMGI. Imaging in the RD-2000N was
done with a 0.30-NA 248-nm excimer laser stepper with
an exposure dose of 230 mJ/cm^2. Transfer of the patterns
into the PMGI was accomplished by deep UV broadband
(220-280 nm) flood exposure with a proximity exposure
tool, the previously imaged RD-2000N acting as a mask
due to its strong absorbance in the deep-UV region.
0.6-micron lines and spaces produced in such a resist
system are shown in Fig. 6.12. Orvek et al. [66] have
investigated PCM systems employing McDermid PR-1024MB
and Shipley XP8843 as the top imaging layers on PMGI as
the planarizing layer.

6.2.6. Diffusion-Enhanced Silylating Resists

As discussed in Sec. 6.2.5, multilevel resist
structures offer various advantages in high-resolution
optical lithography due to their ability to pattern
submicron features with large aspect ratios and thus
effectively overcome the fundamental depth-of-focus
limitations of optical imaging systems. However, there
are also several disadvantages of multilayer processing
due to the cost and complexity of the extra process
steps. Recently, attempts have been made to develop
single-layer resist systems that combine most of the
advantages of multilayer resists with few of their
drawbacks. A step taken in this direction is the DESIRE
(diffusion-enhanced silylating resist) process [119-
122] in which a gas-phase silylating step following
exposure produces an etch barrier in a thin surface
layer of a single-layer planarizing resist. The basic
requirements for such a resist are that it must have a
sufficient concentration of functional groups that can
react with a silylating agent, and it must contain a
sensitizer that can produce an imagewise change in the
diffusivity of the silylating agent into the resist.
An example of such a system is the UCB Plasmask resist
which contains a phenolic resin and a diazoquinone

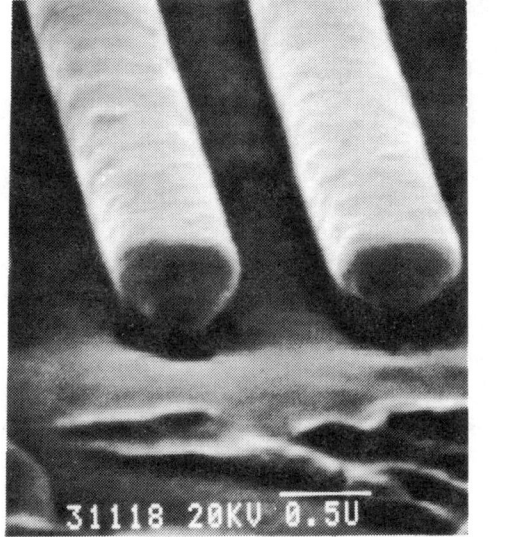

(a)

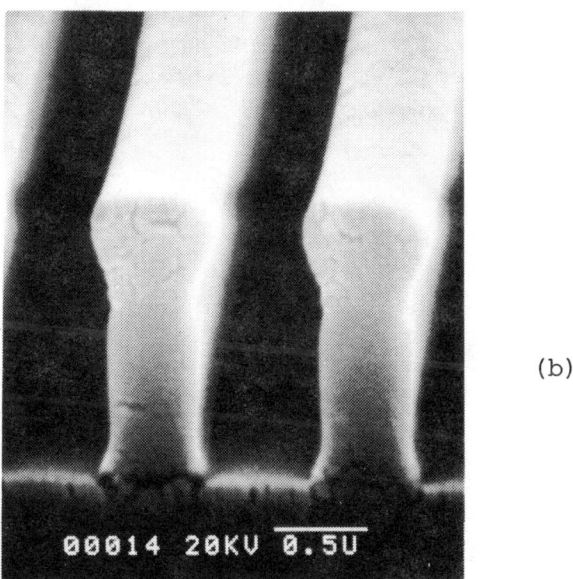

(b)

Fig. 6.12   0.6-micron   projection   images   obtained by portable conformable masking in a structure comprising 0.5-micron-thick   RD-2000N   on top of 1.0-micron-thick PMGI, showing resist profiles in (a) RD-2000N and (b) the PCM structure. [From Ref. 56]

sensitizer [121].

Figure 6.13 depicts the DESIRE process flow. The Plasmask resist is coated to a thickness of ~2 microns and prebaked at 100 °C. After illumination with the exposure tool, the resist is treated with a vaporized silylating agent, such as hexamethyldisilazane (HMDS), at a temperature of 150 °C. This step causes diffusion of the silylating agent into the top 150-250 nm of the resist, but only in the exposed regions, because unexposed diazoquinone acts as a barrier to diffusion of the silylating agent into the phenolic resin. The silicon incorporated into the resist during the above step is chemically bound to the phenolic resin, which gives long-term stability to the silylated wafers. In the final step the wafers are developed in an oxygen plasma that converts the silicon into silicon dioxide, thus creating a thin etch barrier that stops etching in the exposed areas and permits removal of the unexposed resist. Orvek et al. [66] have investigated the DESIRE process for lithography at 248 nm with a 0.35-NA step-and-repeat system; 0.5-micron lines and spaces obtained in 1.2-micron-thick resist are shown in Fig. 6.14.

6.3. RESISTS FOR EXPOSURE AT WAVELENGTHS BELOW 200 NM

Excimer lasers are capable of emitting at several wavelengths below 200 nm, as discussed in Chapter 5. Of the many sub-200-nm wavelengths listed in Table 5.1, two are readily available from commercial excimer laser systems: 193 nm from ArF and 157 nm from F_2. Although there have been some reports on lithography at these wavelengths, no single-layer resist systems have yet evolved that are suitable for lithographic processes with a conventional wet development step. The primary reason for the lack of progress in resist development at 193 and 157 nm is the tendency of most known resist materials to photo-ablate at these wavelengths. Indeed, there has been a large volume of research on ablative processes at sub-200-nm wavelengths, including sub-micron patterning, in a variety of materials; these developments are discussed at length in Chapter 7.

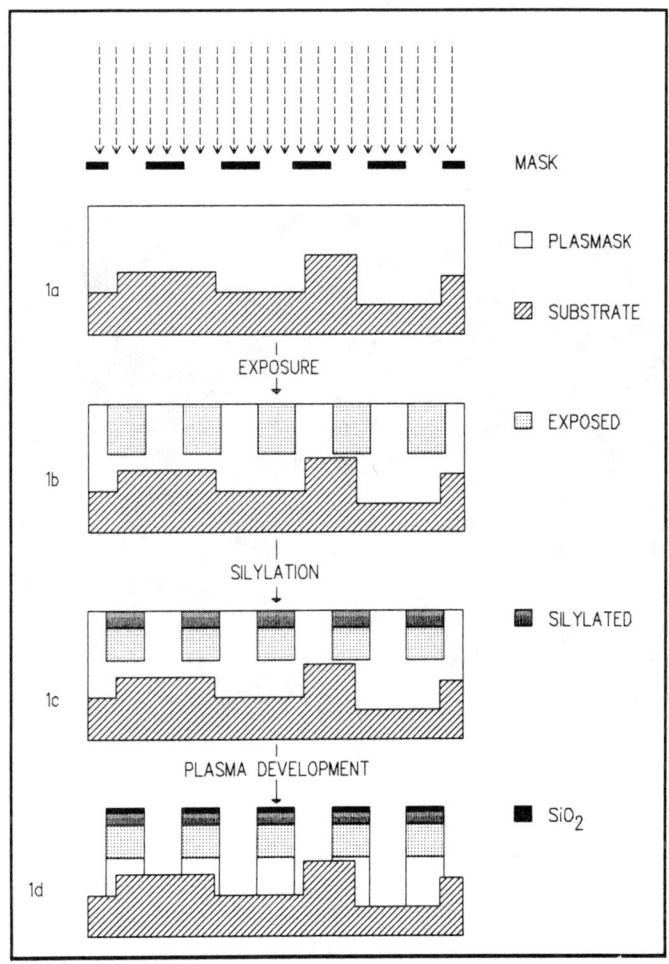

Fig. 6.13   Process flow for the DESIRE resist.   [From Ref. 121]

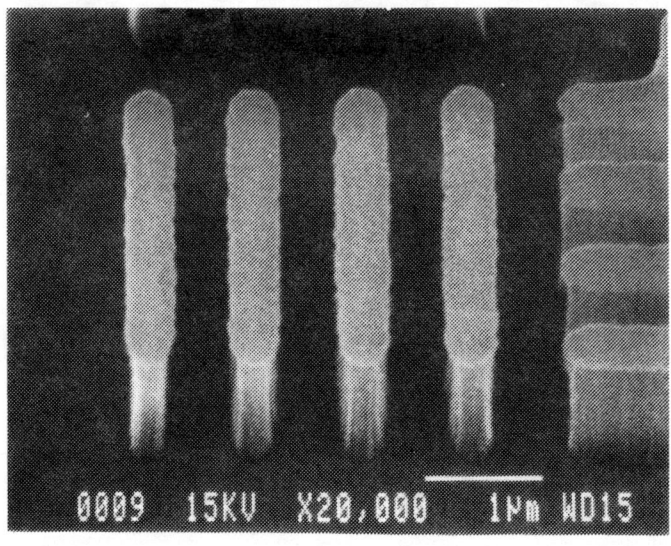

Fig. 6.14   0.5-micron images   obtained   in 1.2-micron-
thick DESIRE resist by projection printing on a 0.35-NA
stepper. [From Ref. 66]

Most resist materials discussed in this chapter also absorb heavily in the spectral region below 200 nm, making conventional lithographic processing difficult.

Exposures at 193 nm have been carried out in PMMA by Cullmann [22]. The resist layer was deposited to a thickness of ~1 micron, and following an exposure dose of >1 J/cm^2, the wafers were developed in pure MIBK. Features down to 0.2 micron in size were resolved in contact printing, illustrating the high-resolution capability of PMMA at this wavelength. An example of images obtained in the above fashion is shown in Fig. 4.4. Several materials other than PMMA have also been investigated for patterning at 193 nm; however, the phenomenon observed in all cases was ablation. These results will be described in the following chapter.

In lithography experiments in the vacuum UV region with the 157-nm F$_2$ excimer laser, Craighead et al. [9, 23] have investigated PMMA, a methacrylate copolymer, and a trilevel resist structure. Experiments with PMMA were performed on a film that was only 0.1 micron in thickness due to the high absorbance of this material at 157 nm. Even then, clean development in the exposed regions was not obtained as a result of the fact that the absorbance of PMMA at 157 nm not only does not bleach, it in fact increases with the exposure dose, as shown in Fig. 6.15(a). Lithographic images in 0.1-micron-thick PMMA exposed at 157 nm with a dose of 0.5 J/cm^2 and developed in a 3:7 solution of cellosolve and methanol for 10 s are shown in Fig. 6.15(b).

A copolymer consisting of 91% methylmethacrylate (MMA) and 9% methacrylic acid (MAA) by volume behaved quite differently from PMMA [9,23]. It also had high absorbance at 157 nm, but unlike PMMA, the copolymer's absorbance was found to bleach with exposure, as shown in Fig. 6.16(a), producing better overall development characteristics. Lithographic patterning in the MMA-MAA copolymer was done using a 0.2-micron-thick film with an exposure dose of 0.2 J/cm^2 and development as above. Figure 6.16(b) shows images obtained in this resist.

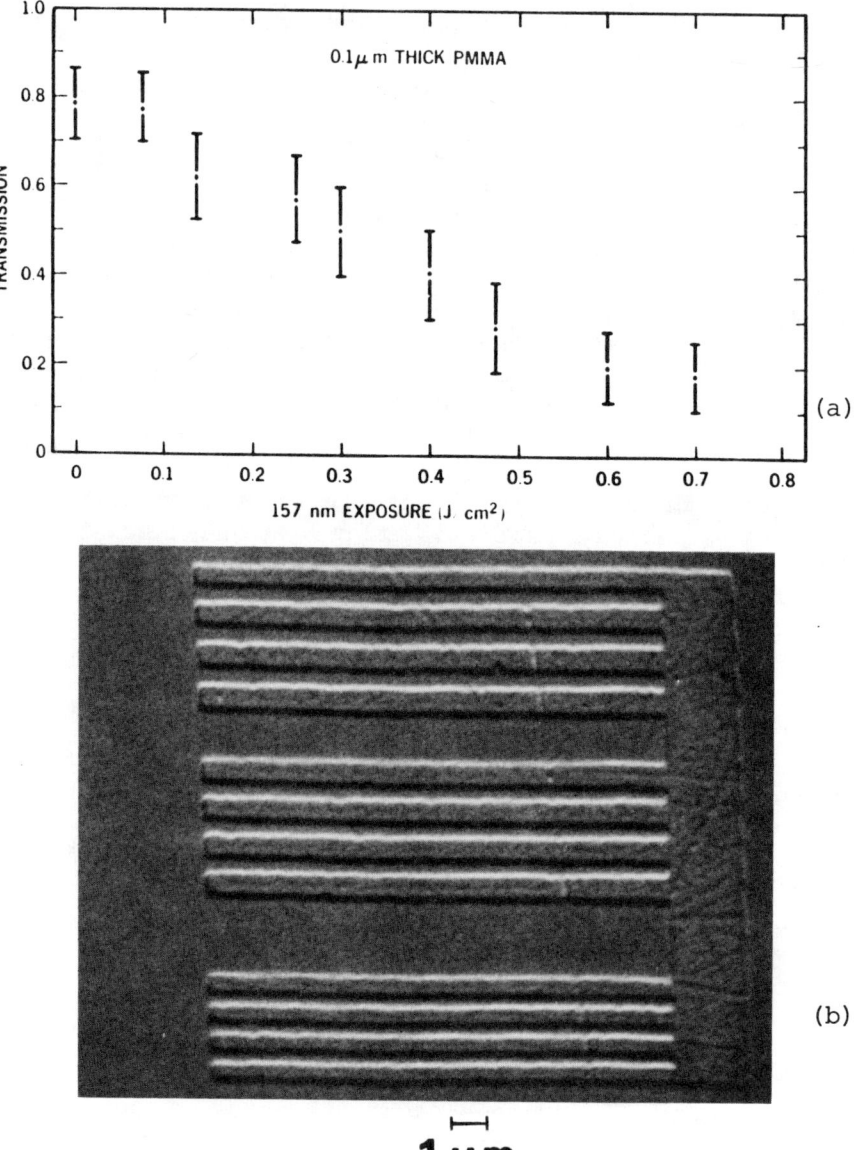

1 µm

Fig. 6.15   (a) Transmission at 157 nm of a 0.1-micron-thick PMMA film as a function of the exposure dose; (b) images obtained in 0.1-micron-thick PMMA by contact printing with a 157-nm excimer laser. [From Ref. 9]

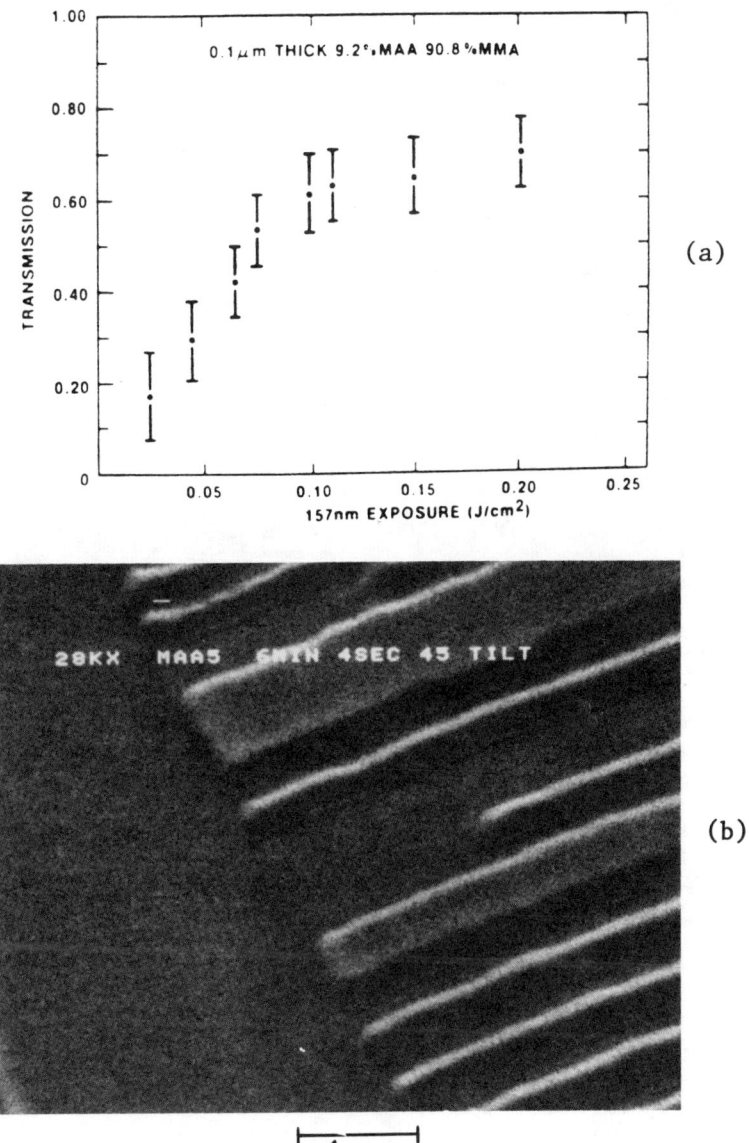

(a)

(b)

Fig. 6.16   (a) Transmission at 157 nm of a 0.1-micron-thick MMA-MAA copolymer film as a function of exposure dose; (b) images obtained in 0.2-micron-thick MMA-MAA copolymer by contact printing with a 157-nm excimer laser. [From Ref. 9]

The trilayer resist system studied by Craighead et al. [9,23] consisted of a 50-nm-thick PMMA layer on top of a 25-nm-thick germanium layer deposited on a 0.25-micron-thick layer of polyimide on silicon. The thin upper layer of PMMA was exposed at 157 nm with a dose of 0.5 J/cm^2 and following development, patterns were transferred by reactive-ion etching, first into the Ge and then into the polyimide. 0.15-micron line features produced with such a process are shown in Fig. 4.6. The mask used for all 157-nm imaging experiments described in this section was made using a trilayer resist system similar to the above configuration except that the PMMA layer was patterned by electron-beam lithography, and the resist structure was fabricated on a CaF_2 substrate rather than on silicon.

6.4. PHOTORESIST RECIPROCITY BEHAVIOR

An important and interesting aspect of the use of excimer lasers for lithography is the effect the short duration (10-20 ns) pulses of high power (10^4-10^7) may have on the photoresist. One might expect that these high-power pulses will produce photochemical changes in the resist different from those produced with lamp illumination, with a concomitant loss of reciprocity in the laser exposures. A photoresist exposure process is said to obey the law of reciprocity if the product of the light intensity and the required exposure time is intensity-independent. This was shown to be the case by Dill et al. [123] for AZ 1350J photoresist for power densities up to 415 mW/cm^2. Using a two-step kinetic model applicable to both positive and negative resists, Albers and Novotny [124] have calculated the intensity dependence of photochemical reaction rates and found that saturation would occur for intensities >5 kW/cm^2. With the large instantaneous power densities ($\sim 10^7$ W/cm^2) involved in the laser exposures, one may expect a significant departure from reciprocity. This can be investigated quantitatively by examining a number of different resist properties - e.g., bleaching, photo-sensitivity, and dissolution rate - at various power levels from the laser. Several reports have appeared in the literature on the characterization of these properties at different excimer laser wavelengths for a variety of resist materials.

Jain et al. have studied the reciprocity behavior of several photoresists using the XeCl excimer laser at 308 nm [1,5,10]. In their bleaching experiments [10], 1-micron-thick films of AZ 2400 and the diazonaphtho-quinone-novolak-based IBM experimental resists ER1 and ER2 were spun on quartz disks. With a fixed laser pulse energy, i.e., for a given peak power, the films were irradiated with different integrated energy doses, i.e., with different numbers of pulses. Without developing the films, their absorbance before and after exposure was measured with a spectrophotometer, as shown in Fig. 6.17. The absorbance value for a characteristic peak (e.g., 402 nm in Fig. 6.17), indicating the amount of photosensitizer present in the film, may now be plotted versus different integrated energy doses, but for the same peak power. A number of such plots obtained for AZ 2400 for different laser peak powers are shown in Fig. 6.18, which also includes a plot for exposures with a mercury lamp. Since the slopes of the various lines are essentially identical, it may be concluded that in the laser power range investigated (0.48-5.4 MW/cm^2), there are no observable reciprocity failure effects if one uses bleaching as a criterion. This can also be seen by picking off from each plot in Fig. 6.18 the dose D_B required for a given level of bleaching and noting, as shown in Fig. 6.19, that D_B is essentially independent of the peak power.

In the photosensitivity experiments carried out by Jain et al. [5], the resist films were exposed as before. The wafers were developed until the exposed region for each integrated dose was completely removed, and then the remaining thickness of the unexposed area was measured. A plot was made of this remaining resist thickness versus the dose. The dose at which the linear portion of this plot extrapolates to a remaining resist thickness of 1 micron was then called D_s, the photo-sensitivity of the resist. Such plots for four peak powers (corresponding to laser pulse energies of 0.08-

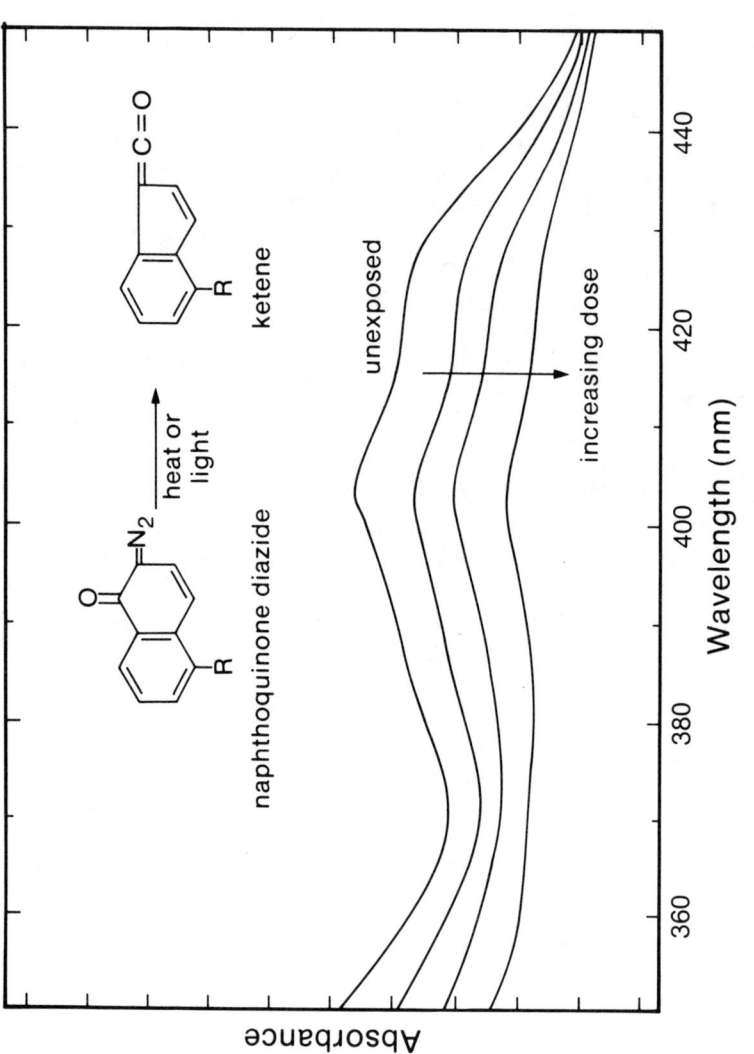

Fig. 6.17 Absorbance spectra of a novolak-based resist following bleaching exposures of various doses from a XeCl excimer laser at 308 nm. [From Ref. 10]

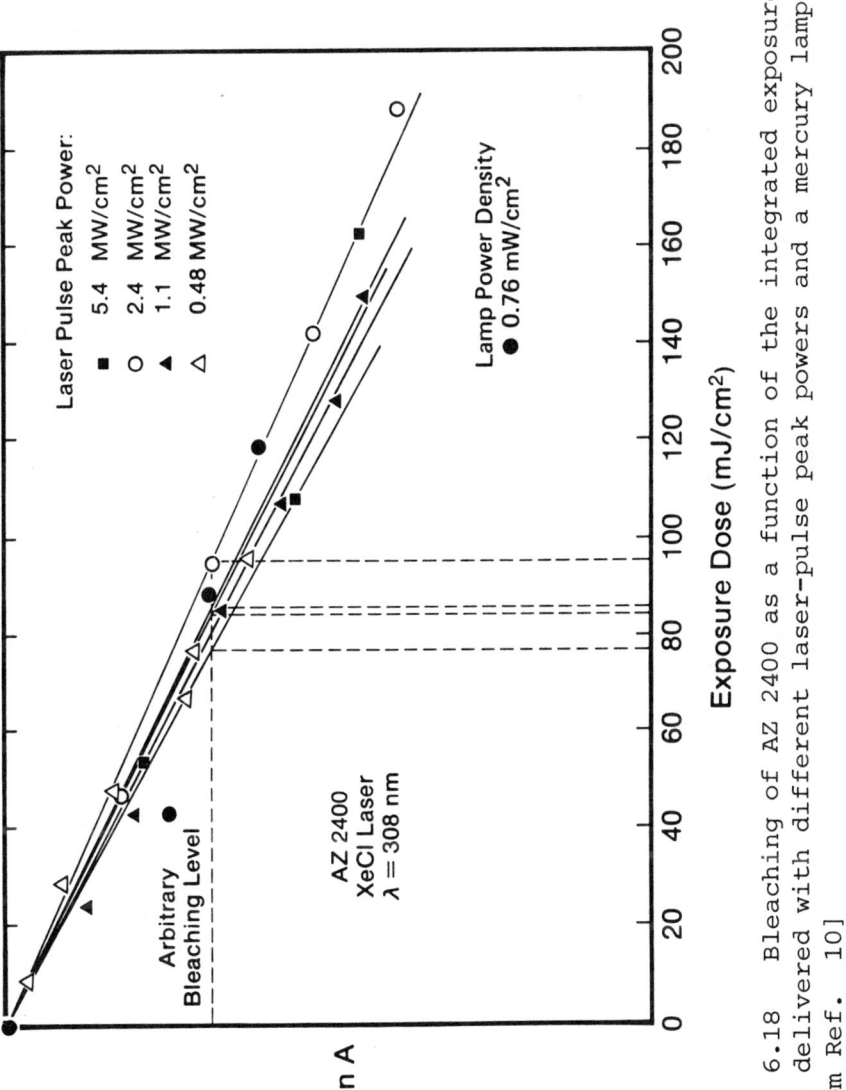

Fig. 6.18 Bleaching of AZ 2400 as a function of the integrated exposure dose delivered with different laser-pulse peak powers and a mercury lamp. [From Ref. 10]

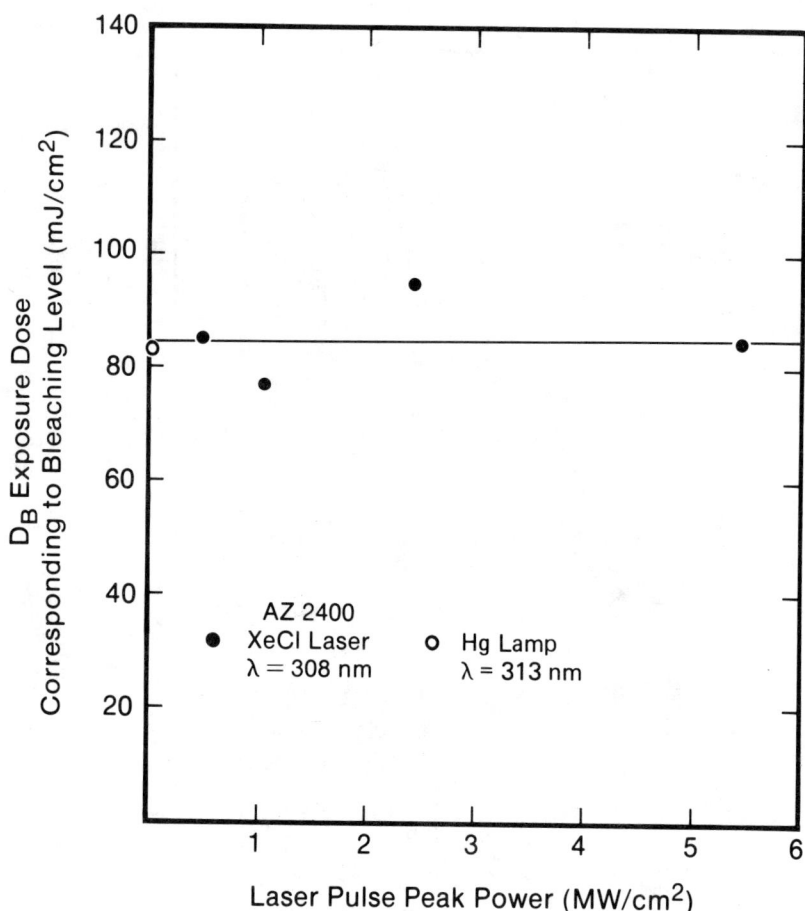

Fig. 6.19   Dependence of the integrated exposure dose corresponding to an arbitrary bleaching level on the peak power from the laser and the mercury lamp, showing absence of reciprocity failure in AZ 2400 resist. [From Ref. 10]

3.9 mJ/cm^2) for AZ 1450J photoresist are shown in Fig. 6.20. It is seen that for all power levels, D_s is in the 30-60 mJ/cm^2 range, comparable to the sensitivity observed with mercury lamp exposure for this resist, within a factor of 2. Thus, in the power range of 8-390 kW/cm^2, there is little or no reciprocity failure in AZ 1450J when photosensitivity, as described above, is used as the criterion. The photosensitivities of three other mid-UV resists - AZ 2400 and the IBM resists ER1 and ER2 - have also been measured at the 308-nm XeCl excimer laser wavelength [1]. The results, shown in Fig. 6.21, indicate that for these resists also, the reciprocity failure, if any, is limited to within a factor of 2 to 3.

Reciprocity effects may also be investigated by measuring the dissolution rate as a function of dose delivered with pulses of different peak powers. Results obtained by Kawamura et al. [4,112] in AZ 2400, PMMAB (PMMA with 10% benzoin), and PMMA are shown in Figs. 6.22(a), 6.22(b), and 6.23, respectively. Note that whereas the dissolution characteristics of AZ 2400 are essentially independent of the incident peak power, the behavior of both PMMAB and PMMA is marked with strong intensity dependence. For example, in PMMA (Fig. 6.23), the development rate for an integrated exposure dose of 700 mJ/cm^2 delivered with pulses of 5.6 MW/cm^2 peak power is eight times larger (2.4 micron/min versus 0.3 micron/min) than for the same dose delivered with 1.3-MW/cm^2-peak-power pulses.

Reciprocity behavior of a methacrylate polymer has also been investigated by bleaching experiments. In PMMA films doped with 20% acridine by weight, Sheats has observed strong intensity-dependent photobleaching in exposures with a 248-nm KrF laser [14]. In these studies, transmission of 248-nm laser pulses through a 1-micron-thick sample film deposited on a 0.1-mm-thick quartz disk was measured after illumination with prior bleaching pulses. Figure 6.24 shows the transmission of low-intensity (<2 MW/cm^2) pulses following bleaching with a single pulse of different intensities. Note the highly nonlinear intensity-dependence of the bleaching;

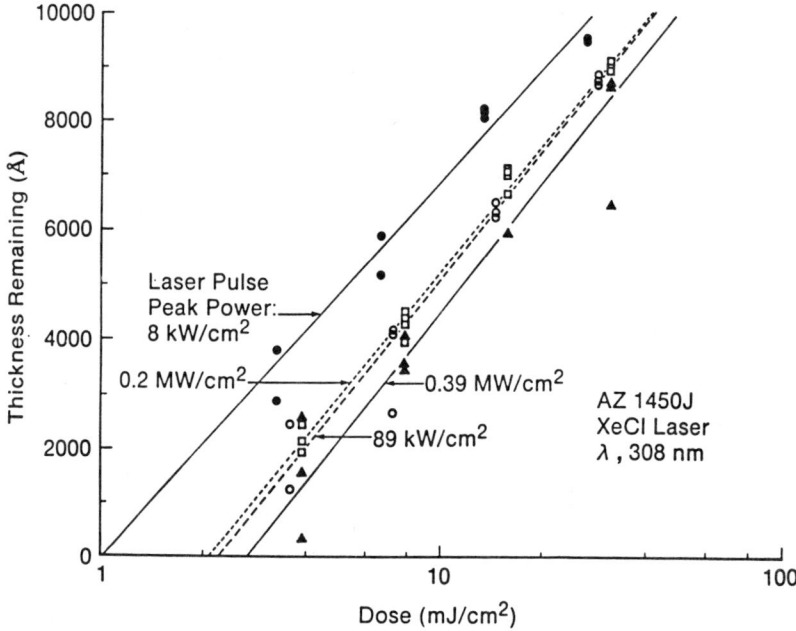

Fig. 6.20 Photosensitivity of AZ 1450J resist for
308-nm excimer laser exposures - obtained by plotting
the remaining thickness of the unexposed resist versus
the integrated dose - as a function of the laser pulse
peak power, showing absence of reciprocity failure.
[From Ref. 5]

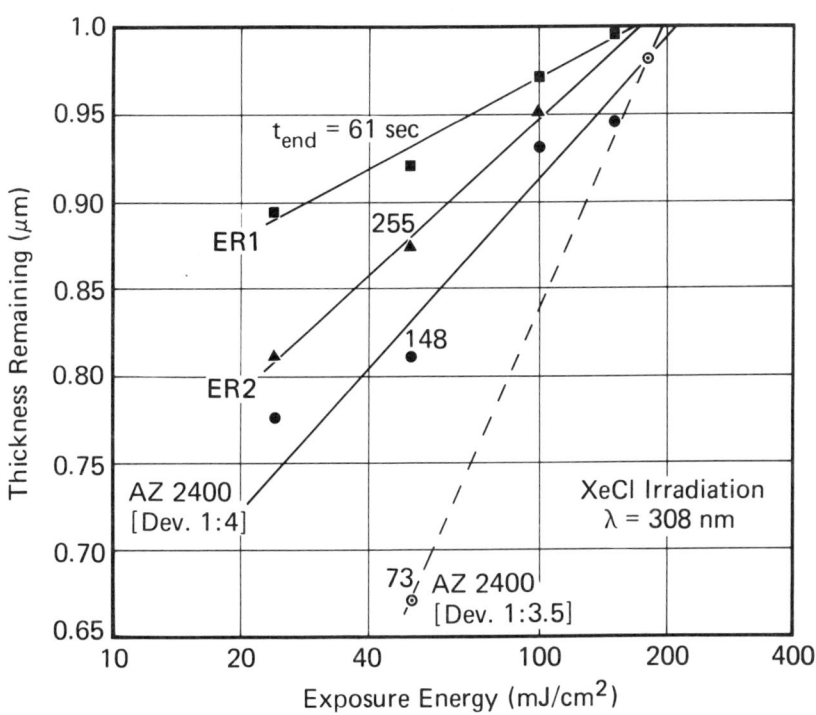

Fig. 6.21   Photosensitivity of various mid-UV resists for 308-nm XeCl excimer laser irradiation obtained by plotting the remaining unexposed resist thickness versus integrated dose. The endpoint development time with 50 mJ/cm² exposure is also shown. [From Ref. 1]

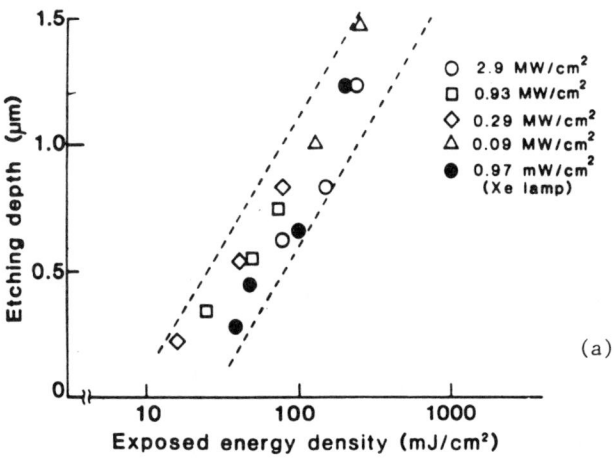

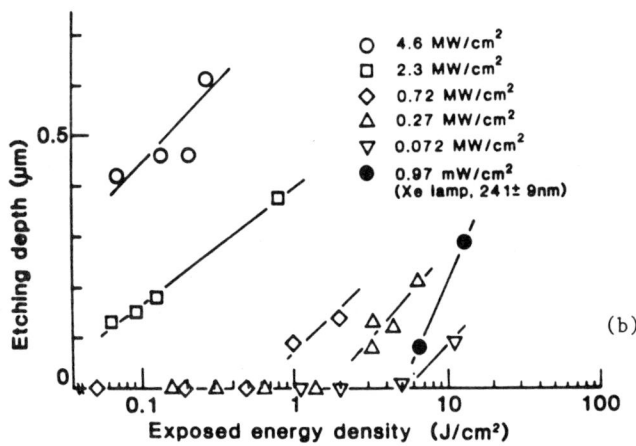

Fig. 6.22   (a) Dissolution rate of AZ 2400 for 308-nm
XeCl excimer laser irradiation with various peak power
pulses, showing absence of reciprocity failure; (b)
dissolution rate of PMMAB for 248-nm KrF excimer laser
irradiation with various peak power pulses, showing
severe reciprocity failure. [From Ref. 112]

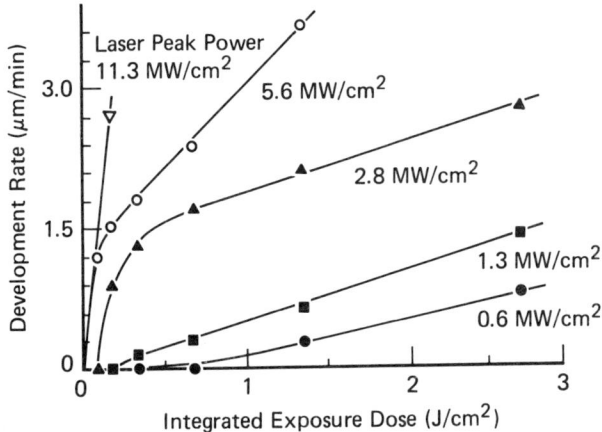

Fig. 6.23   Dissolution rate of PMMA for 248-nm KrF laser irradiation with various peak-power pulses, showing severe reciprocity failure. [From Ref. 4]

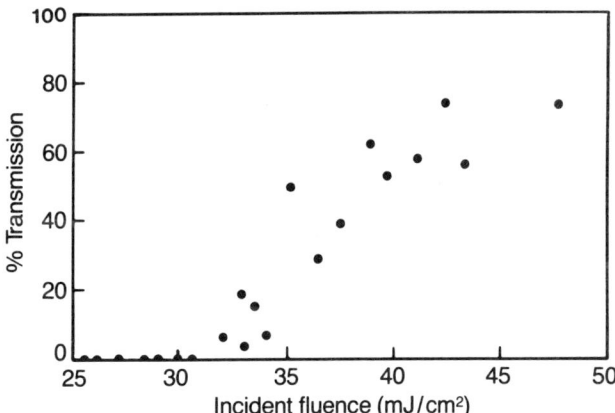

Fig. 6.24   Transmission of a PMMA-acridine film for low-intensity 248-nm pulses versus fluence of a single prior bleaching pulse at 248 nm from an excimer laser, showing highly nonlinear intensity-dependent bleaching. [From Ref. 14]

for example, films exposed with a 30 mJ/cm^2 pulse show no bleaching (T = 0), whereas exposure with a 40 mJ/cm^2 bleaching pulse results in 60% transmission.

Abe et al. have studied the reciprocity behavior of PMMA in irradiation at 222 nm with a KrCl excimer laser [125]. 1-micron PMMA films spun on silicon wafers were given various exposure doses using laser pulses of different energies. Following exposures the wafers were developed in methyl isobutylketone (MIBK) for 6 min and the remaining resist thickness measured. As in the case of the reciprocity behavior of PMMA in exposures at 248 nm [4], described above, the development rate for 222-nm exposures was also found to increase in a highly nonlinear manner with the pulse energy, as shown in Fig. 6.25(a). This may be viewed as an effective increase in sensitivity at higher pulse energies, and has been attributed to heating of the resist. Abe et al. have employed a resist sensitivity model that incorporates an activation energy for the chain scissioning photo-chemical reaction, and obtained good agreement between the observed dissolution rates and those calculated from the model, as shown in Fig. 6.25(b).

The contribution of thermal effects in reciprocity failure has also been suggested in 248-nm KrF laser exposures of the inorganic resist $Ag_2Se/GeSe_2$ [15,19]. In this study, it was found (Fig. 6.26) that, for pulse energies up to 5.2 mJ/cm^2, the dose required for full development was intensity-independent, and constant at 130 mJ/cm^2. However, for pulse energies >5.2 mJ/cm^2, heating of the resist causes the Ag_2Se to undergo a phase transition to a high-temperature (>130 °C) phase, from which diffusion of Ag ions into $GeSe_2$ is greater, producing a larger differential between the etch rates of exposed and unexposed regions.

It is clear, from the extensive studies described above, that the reciprocity behavior of photoresists on excimer laser irradiation varies from one material to another, and also depends on the intensity range investigated. It may be said, however, that for most

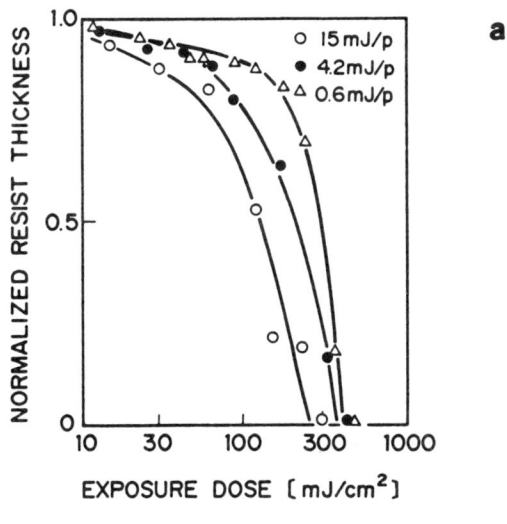

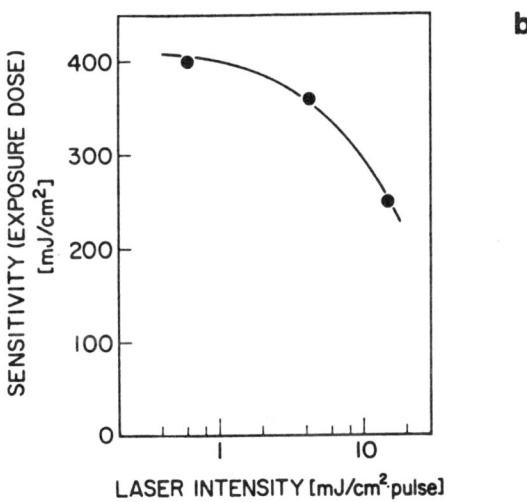

Fig. 6.25   (a) Photosensitivity of PMMA for 222-nm
KrCl excimer laser exposures with different laser pulse
energies; (b) dependence of resist sensitivity on laser
pulse intensity; the dots are experimental values for
PMMA and the curve represents a theoretical calculation
based on the assumption that resist heating causes
reciprocity failure. [From Ref. 125]

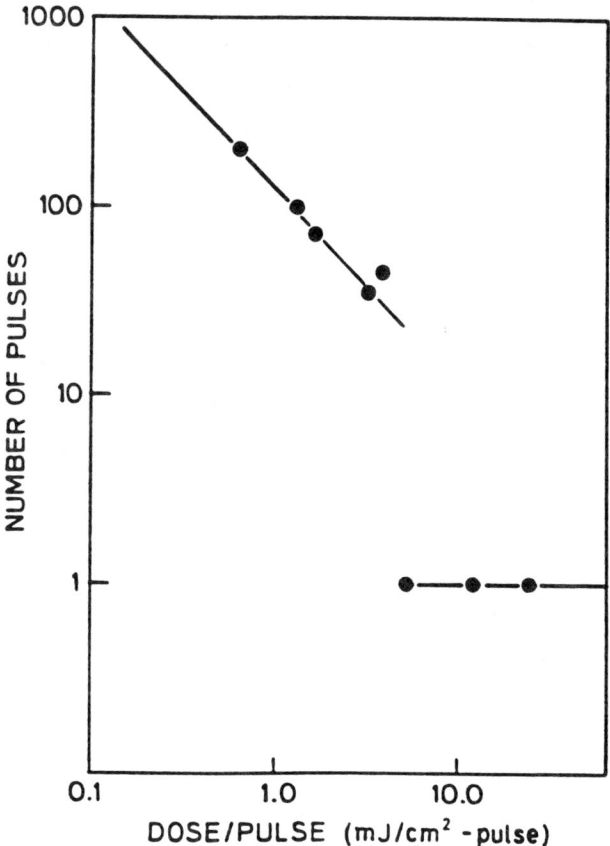

Fig. 6.26   Intensity-dependence of photosensitivity of
the inorganic resist $Ag_2Se/GeSe_2$ in 248-nm KrF excimer
laser exposures, showing onset of a reciprocity failure
mechanism at a pulse energy of ~ 5.2 mJ/cm^2. [From Ref.
15, © 1984 IEEE]

conventional mid-UV resists, there is no reciprocity
failure at exposure doses of lithographic interest, and
all of the higher flux available from excimer lasers
can be used to reduce the required exposure times. In
deep UV materials, although significant intensity-
dependent nonlinearities have been observed in various
properties, it is surprising that in spite of the 10^8
times higher instantaneous power levels used in the
laser exposures compared to the mercury lamp case, the
nonlinearities are so small. It is unlikely, in our
view, that the photochemical processes taking place in
the two cases are the same. Turro et al. [126] have
observed that in exposure of diphenyldiazomethane
and tetraphenyloxirane with deep UV light, the photoproducts produced by KrF laser excitation are different
from those produced by conventional lamp excitation.
For materials of lithographic interest, we expect that
further investigations, including careful analysis
of reaction products to determine how generation of
photoproducts depends on intensity, should provide
valuable insight into understanding their reciprocity
behavior.

7. Excimer Laser Etching and Deposition

In addition to the wide range of excimer laser applications in microlithography in which a positive or negative photoresist is exposed and then developed, there is also interest in several other techniques for patterning with excimer lasers, including etching and deposition. Such techniques are attractive because they make it possible to remove or deposit material selectively at localized areas on a sample, leading to practical applications such as repair of open and short defects on masks, etching via holes in polymer films used in circuit boards, tissue removal in various medical applications, etc. These materials processing phenomena have been explored with several excimer laser wavelengths in a wide variety of media. Although a comprehensive review of all excimer laser etching and deposition phenomena and applications is not possible within the scope of this book, in this chapter we present selected examples of such processes along with a discussion of the basic underlying phenomena.

One of the most extensively studied excimer laser etching phenomena is ablative photodecomposition of organic materials, first investigated by Srinivasan et al. [127,128], who etched 5-micron features in poly-(ethylene terephthalate) films with an ArF excimer laser at 193 nm. In this type of process, it is found that a number of organic materials, upon irradiation with intense UV and vacuum UV (VUV) light, undergo a photoetching mechanism in which material removal takes place at the molecular layer level. The phenomenon appears to be characterized by absorption of almost all the incident radiation in a very thin upper layer of the material, followed by breaking of a large number of bonds and chains (whose energies are less than the energy of the incident photons), and finally ejection of the photo-products as small molecules. A simplistic illustration of the mechanism is shown in Fig. 7.1. There also seems to be an intensity threshold in the vicinity of 10-50 mJ/cm^2 below which little material removal is observed [129]. Although the above bond-

176

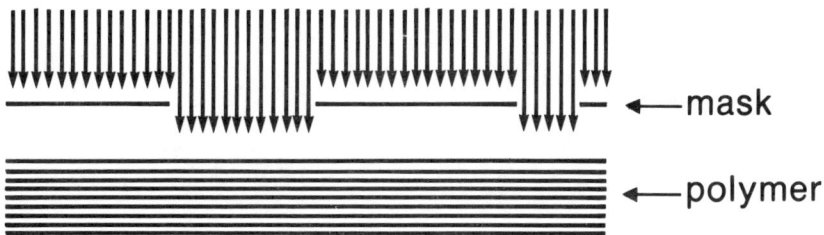

(a) Light Absorption

(b) Bond Breaking

(c) Ablation

Fig. 7.1   Simplified illustration of the mechanism of
ablative photodecomposition, showing light absorption,
bond breaking, and ejection of photoproducts. [From
Ref. 129]

breaking process has been widely described as the
main mechanism underlying ablative excimer laser
etching [127-129], there is also evidence that thermal
effects contribute to the process, especially in
irradiation with wavelengths >200 nm [130-133].

Submicron patterning with direct high-resolution
excimer laser photoetching has been demonstrated by
Rice and Jain in various resists and polymers using 193
nm contact exposures with an ArF laser [12]. Table 7.1
summarizes the materials investigated by these authors,
the experimental conditions, and the results obtained.
Note that some of the media were 1-micron-thick films
deposited on silicon wafers, whereas others were in the
form of free-standing thin sheets. The samples were
irradiated with laser pulses of 13 mJ/cm^2 energy, with
the total dose required to etch through a thickness of
1 micron being in the vicinity of 1 J/cm^2 for most
materials. Figure 7.2 shows 1-micron and 0.75-micron
lines produced by etching in poly(methyl isopropenyl
ketone) (PMIPK). A 0.5-micron line etched in the diazonaphthoquinone-novolak resist ER1 is shown in Fig. 7.3,
in which clean etching and a 3:1 aspect ratio may be
noted. An example of submicron patterning by excimer
laser etching in AZ 2400 has been previously given in
Fig. 4.5. It is observed that some of the materials,
e.g., PMIPK, demonstrate very clean etching with little
debris, while others show the presence of a considerable
amount of particulate matter near the etched features.
Although the set of specific photochemical and thermal
processes responsible for etching may vary from one
material to another, ablative photodecomposition may be
considered a primary mechanism in each, in which case
the differences in the etching behavior in different
media may be attributed to variation in the fraction of
nonvolatile photofragments produced in the process. For
example, in the case of PMIPK (Fig. 7.2), the clean
etching suggests that there are practically no nonvolatile compounds formed, whereas the particulates
seen in the vicinity of the etched region in ER1 (Fig.
7.3) and AZ 2400 (Fig. 4.5) point to a significant
contribution of nonvolatile matter in the generated
photoproducts. It is also possible that some of the

Table 7.1 Polymers and resists directly photoetched with 193-nm ArF excimer laser. [From Ref. 12]

Material	Sample Configuration	Processing Prior to Exposure (°C/min)	Minimum Linewidth Etched (μm)	Etched Wall Angle (°)	Comments
PMIPK	1-μm-thick film on Si wafer	120/20	0.5	60	Clean etching; very little debris
AZ 2400	1-μm-thick film on Si wafer	90/20	0.3	72	Considerable amount of debris
Novolak-based resist ER1	1-μm-thick film on Si wafer	90/15	0.5	75	Clean etching; debris in unexposed areas
PMMA	1-μm-thick film on Si wafer	180/60	0.5	–	Clean etching; little debris
PMMA	1.25-mm-thick sheet	–	0.8	–	Some debris
Polyester	125-μm-thick sheet	–	1.0	–	Some debris
Polyimide	75-μm-thick sheet	–	1.0	–	Scattered particulate deposits
Cellulose Acetate	125-μm-thick sheet	–	1.0	–	Considerable amount of debris; indication of thermal damage
Polycarbonate	12.5-μm-thick sheet	–	1.5	–	Considerable amount of debris; indication of thermal damage

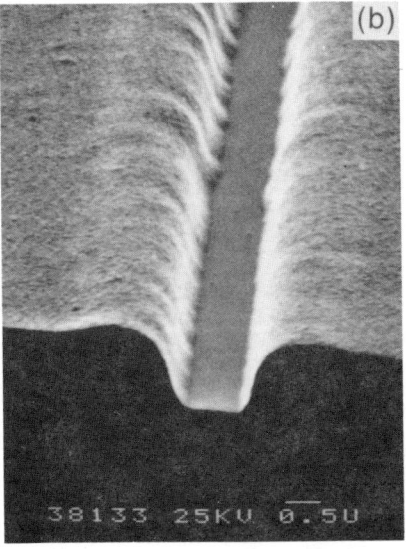

Fig. 7.2   Direct photoetching in 1-micron-thick PMIPK
with a 193-nm ArF excimer laser, showing feature sizes
of (a) 1 micron and (b) 0.75 micron. The total exposure
dose was 1 J/cm^2. [From Ref. 12]

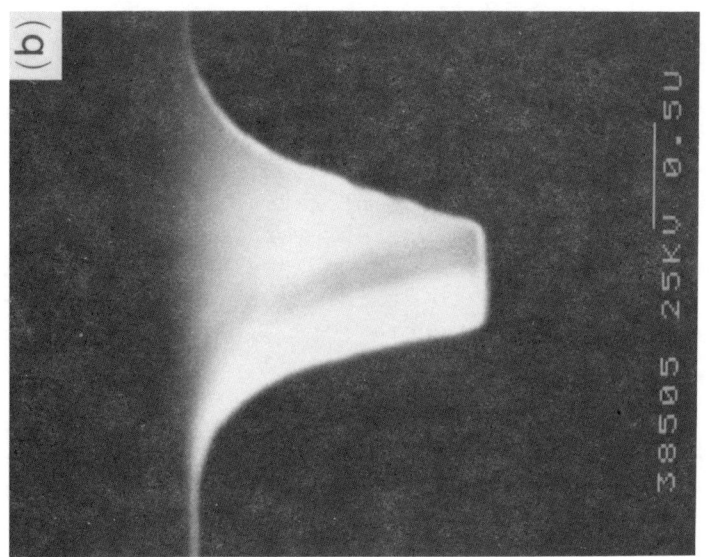

Fig. 7.3   A 0.5-micron line directly photoetched in a diazonaphthoquinone-novolak photoresist with a 193-nm ArF excimer laser. The total exposure dose was 1 J/cm². Note the clean etching, the 3:1 aspect ratio, and some debris in the unexposed regions. [From Ref. 12]

ejected volatile matter may be redepositing on the
sample surface following exposure with the laser pulse.

Excimer laser ablative photoetching of organic
materials has also been carried out with a vacuum UV F_2
laser at 157 nm. Henderson et al. [23] have etched 0.2-
micron features in nitrocellulose by contact exposure
with 0.1 J/cm^2 pulses through a stencil mask. At this
pulse energy, an etching rate of 100 nm per pulse was
observed, which is similar to the rate of 1 micron per
J/cm^2 observed with 193-nm exposure of several organic
materials discussed above. Using lower energy pulses,
nitrocellulose was found to exhibit a threshold for onset of etching at a pulse energy of 25 mJ/cm^2. Dyer and
Sidhu [30] have investigated 157-nm F_2 laser etching of
poly(ethylene terephthalate), polyimide, and polyethylene. These experiments were performed on 50-micron-
thick sheets of the materials, and multi-micron-size
features were patterned by contact illumination through
a copper grid mask. The threshold energies for ablative
etching in poly(ethylene terephthalate), polyimide, and
polyethylene were found to be, respectively, 29, 36,
and 67 mJ/cm^2, comparable to the threshold values
observed for organic materials in other investigations.

Although direct excimer laser etching of materials
is of interest as a dry lithographic process, contact
printing as a patterning method is clearly impractical
due to physical ejection of matter and deposition of
debris that take place in the etching process. Thus, it
would be highly desirable to carry out the ablative
exposure of the substrate through an optical projection
system. Latta et al. [17] have demonstrated excimer
laser projection photoetching in various materials with
the 193-nm ArF laser through a projection lens of their
own design. The f/8.3 all-quartz 2:1 reduction lens is
shown in Fig. 7.4. Exposures in two novolak-based
photoresists and polyimide were carried out using 18
mJ/cm^2 pulses, with an integrated dose of 0.9 J/cm^2 for
an etch depth of 1 micron in the patterns. Examples of
projection-etched images obtained by these authors are
shown in Fig. 7.5. Goodall et al. have produced submicron features in PMMA by excimer laser projection

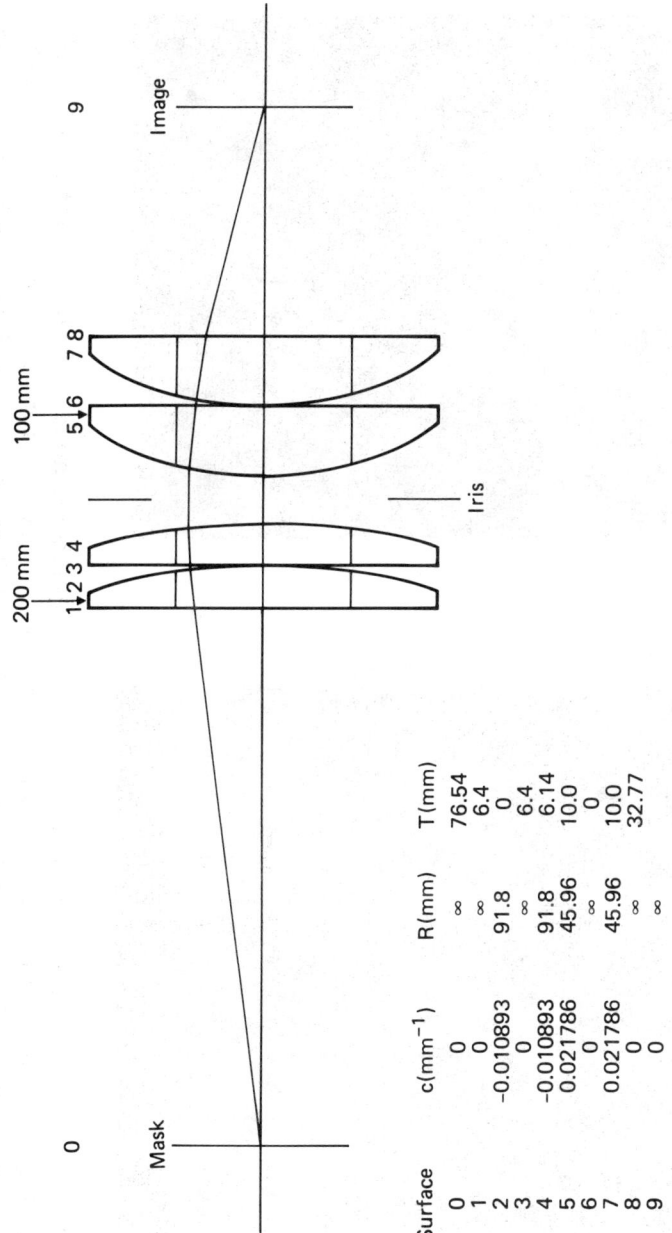

Fig. 7.4 An f/8.3 all-quartz 2:1 reduction lens designed for projection photoetching with excimer lasers. The curvature c, radius R, and thickness T are indicated for each surface. [From Ref. 17]

Surface	c(mm⁻¹)	R(mm)	T(mm)
0	0	∞	76.54
1	0	∞	6.4
2	-0.010893	91.8	0
3	0	∞	6.4
4	-0.010893	91.8	6.14
5	0.021786	45.96	10.0
6	0	∞	0
7	0.021786	45.96	10.0
8	0	∞	32.77
9	0	∞	

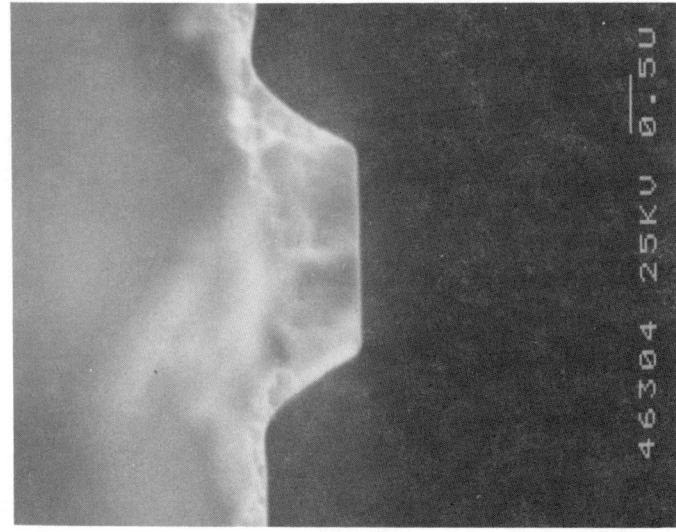

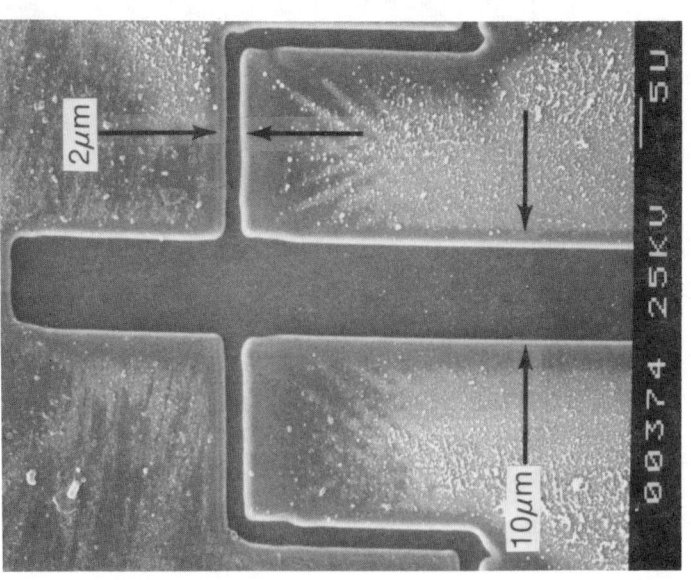

Fig. 7.5   Patterns etched in a diazonaphthoquinone-novolak photoresist by 193-nm ArF excimer laser imaging through a projection lens and ablation with a total dose of 900 mJ/cm². (a) 10- and 2-micron-wide lines and (b) cross section of a 2-micron-wide line. [From Ref. 17]

etching through a reduction lens [28].

Excimer laser photo-ablation can also be used to etch various metals and other inorganic materials. Non-chemical ablative photoetching of Ag, Au, Cu, Al, Ni and Cr with the 308-nm XeCl laser has been investigated by Andrew et al. [134]. The metals were deposited on various substrates in film thicknesses of 50-120 nm and ablated with laser pulses 7-ns wide. As shown in Table 7.2, the single-pulse film-removal threshold was found to range between 30 mJ/cm^2 for a 50-nm-thick Au film deposited on polyimide to 240 mJ/cm^2 for an 80-nm-thick Cr film deposited on quartz. Other materials etched by ablation with excimer lasers include diamond and hard carbon films [135].

Another class of excimer laser etching phenomena consists in initiating or enhancing chemical reactions

Table 7.2   Metal films etched by 308-nm
excimer laser photo-ablation. [From Ref. 134]

Metal	Film Thickness (nm)	Substrate	Threshold Fluence (mJ/cm^2)
Ag	100	Mylar	50
Au	50	Kapton	30
Cu	100	Glass	80
Al	120	Mylar	140
Al	120	Glass	200
Ni	120	Perspex	240
Cr	80	Quartz	240

near metal, semiconductor, and insulator surfaces in the presence of the laser photon. These processes involve photodissociation of a molecular gas or liquid in the vicinity of the solid-fluid interface such that a photofragment produced in the dissociation process chemically reacts with the solid. Since the laser-assisted reaction takes place only in the illuminated regions, etching of the solid surface can be pattern-wise localized. In most cases, the etchant gas contains a halogen atom, which is released in the fragmentation step and which then attacks the sample. For example, when silicon is irradiated with XeCl laser pulses in a Cl_2 environment, the Cl_2 molecule photodissociates into Cl atoms, which then react with the Si to form silicon tetrachloride, a volatile compound that can be pumped away from the etched Si surface. Many other materials that have been etched by excimer-laser-induced photo-chemical processes are listed in Table 7.3 along with the etching fluids and the laser wavelengths used.

Table 7.3   Partial list of materials etched by excimer laser photochemical processing.

Material Etched	Etching Fluid	Excimer Laser λ (nm)	Reference
Si	Cl_2	308	[136]
SiO_2	CCl_2F_2	248	[136]
Si	COF_2	193	[136]
Pyrex	H_2	193	[18]
GaAs	CH_3Br, CH_3Cl	193	[136]
GaAs	HBr, HCl	193	[137,138]
W	COF_2	193	[136]

The same basic concept that is employed in photochemical etching for material removal can also be used
for excimer-laser-induced photochemical deposition of
films. As in etching, the laser photons dissociate one
or more molecular gases and produce various photofragments near a solid surface. However, instead of one of
the fragments reacting with the solid and producing a
volatile compound resulting in material removal, in
the deposition process the fragment is adsorbed on the
solid surface, leading to film buildup. In some cases,
the dissociation step may produce two fragments from
two different molecular gases, which may then react
with each other near the substrate surface, resulting
in deposition of a compound material. As an example of
deposition of a metal, a 193- or 248-nm laser photon
can dissociate trimethyl aluminum, $Al(CH_3)_3$, freeing an
Al atom, which then deposits on a sample surface placed
in close proximity to the dissociating molecule. A
compound such as SiO_2 can be deposited by a reaction
between Si and O atoms released by photodissociation of
SiH_4 and N_2O, respectively. This basic process has been
used for excimer-laser-assisted deposition of a wide
variety of materials, including metals, semiconductors,
dielectrics, oxides and refractory materials. Several
other examples of films thus deposited are given in
Table 7.4.

Although photochemical and photothermal etching
and deposition processes using excimer lasers have led
to various novel materials processing techniques, it is
useful to take a critical look at the reaction rates of
these phenomena so as to put their potential industrial
applications in the proper perspective. For example, in
ablative etching of photoresists and polymers, the dose
requirement for removal of a 1-micron-thick film, as
mentioned earlier in this chapter, is $\sim$1 J/cm^2 . Such a
dose is $\sim$20 times larger than the typical requirement
of $\sim$50 mJ/cm^2 for exposing and wet-developing various
mid-UV resists, and $\sim$50-100 times larger than the dose
sensitivity of the single-layer chemical-amplification
deep UV resists described in Sec. 6.2.4. Thus, on the
one hand, excimer laser etching of organic materials
is of interest in patterning those materials that are

Table 7.4   Partial list of materials deposited by
excimer laser photochemical processing.

Material Deposited	Parent Molecule(s)	Excimer Laser λ (nm)	Reference
Al	$Al(CH_3)_3$	193,248	[139,140]
Cr	$Cr(CO)_6$	248	[141]
Si	SiH_4	248	[142]
Ge	GeH_4	193,248	[143]
SiO_2	SiH_4+N_2O	193	[141,144]
Si_3N_4	SiH_4+NH_3	193	[141]
Al_2O_3	$Al(CH_3)_3+N_2O$	193,248	[139]
W	$W(CO)_6$, WF_6+H_2	248,193	[141,145]

difficult to process by conventional methods, and in low-volume situations where the rate of processing is not an important criterion; but on the other hand, it is not attractive as a dry lithographic patterning alternative to conventional lithographic and etching methods.

The reaction rates for photochemical deposition of films are even lower than for photo-ablative etching. For a variety of materials, some of which are listed in Table 7.4, deposition rates of $\sim 10^2$-10^4 nm/min have been found for laser average-power densities of ~ 10 W/cm^2 on the substrate. Such rates correspond to a dose requirement of $\sim 10^2$-10^4 J/cm^2 for deposition of a 1-micron-thick film. It is thus clear that, whereas photochemical deposition processes are of interest in applications such as mask repair and deposition of

special materials, arguments are not persuasive for
viewing them as alternative techniques to conventional
lithographic patterning of a resist followed by film
deposition by methods such as sputtering, evaporation,
etc. Finally, typical rates for material removal by
photochemical etching are less than photochemical film
deposition rates by at least an order of magnitude and,
therefore, considerations similar to those discussed
above apply even more strongly.

8. Outlook

As semiconductor microelectronic devices continue their progress toward finer and finer geometries, optical microlithography not only stubbornly refuses to die, it continues to dominate comfortably alternative technologies. This is primarily a consequence of the emergence of excimer laser lithography, which has made it possible to push the state of the art in optical lithography tooling technology to the half-and sub-half-micron regimes. Simultaneously, progress in deep ultraviolet photoresist technology has provided the required advances in the processing arena.

It is expected that excimer laser lithography systems will continue to enhance their capabilities with advances in all three of their key subsystems: illumination, projection, and alignment. Progress in the projection subsystem is evidenced by the appearance of all-quartz lenses with higher numerical apertures and larger exposure field sizes. However, to exploit fully the potential of excimer laser lithography with wavelengths shorter than 200 nm, advances in lens design and fabrication with materials other than quartz will be required. In the illumination subsystem, progress in excimer laser technology has resulted in the availability of short-wavelength light sources with attractive spectral characteristics, power output, and reliability; further advances in excimer lasers on all these fronts are certain. New developments in alignment techniques have come about at a somewhat slower pace than in other components of the overall lithography system, and it may be that the ultimate limitation in lithography systems will be imposed by their alignment performance rather than image resolution.

Advances in deep ultraviolet resists for excimer laser lithography promise to provide new single-layer materials with high dose sensitivities. Further work in improving their process stability will enhance their practical utility. Simultaneously, progress has been made in various multilayer photoresist systems; these

190

efforts are of key importance in that, as the depth of focus delivered by excimer laser lithography tools decreases with shorter imaging wavelengths, multilayer processing will become more and more attractive, and necessary.

Assessing the impact of the above advances in aggregate, we expect that the dominance of optical lithography witnessed in the last two decades will continue in the 1990s with excimer laser lithography. It is now accepted by consensus in the semiconductor industry that optical tools will predominate in the fabrication of half-micron devices. In the opinion of this author, not only will optical lithography be the primary technology in the 0.35-micron device era, it will also play a significant role in advancing device geometries to the 0.25-micron regime.

References

1. K. Jain, C. G. Willson and B. J. Lin, "Ultrafast deep UV lithography with excimer lasers," IEEE Electron Device Lett., Vol. EDL-3, 53(1982).

2. K. Jain, C. G. Willson and B. J. Lin, "Ultrafast high-resolution contact lithography with excimer lasers," IBM J. Res. Dev. Vol. 26, 151(1982).

3. K. Jain, C. G. Willson and B. J. Lin, "Ultrafast high resolution contact lithography with excimer lasers," Proc. SPIE, Vol. 334, 259(1982).

4. Y. Kawamura, K. Toyoda and S. Namba, "Effective deep ultraviolet photoetching of polymethyl methacrylate by an excimer laser," Appl. Phys. Lett., Vol. 40, 374(1982).

5. K. Jain et al., "Ultrafast deep UV lithography with excimer lasers," Proc. Int. Conf. Microcircuit Engineering, Grenoble, France, 1982, p. 69.

6. K. Jain, "Sources for UV Lithography," Proc. Int. Conf. Microcircuit Engineering, Grenoble, France, 1982, p. 293.

7. G. M. Dubroeucq and D. Zahorski, "KrF excimer laser as a future deep UV source for projection printing," Proc. Int. Conf. Microcircuit Engineering, Grenoble, France, 1982, p. 73.

8. K. Jain, "Laser applications in semiconductor microlithography," Lasers and Applications, Vol. II, No. 9, Sept. 1983, p. 49.

9. H. G. Craighead et al., "Contact lithography at 157 nm with an F_2 excimer laser," J. Vac. Sci. Tech. B, Vol. 1, 1186(1983); see also J. C. White et al., "Submicron vacuum ultraviolet contact lithography with an F_2 excimer laser," Appl. Phys. Lett., Vol. 44, 22(1984).

10. S. Rice and K. Jain, "Reciprocity behavior of photoresists in excimer laser lithography," IEEE Trans. Electron Devices, Vol. ED-31, 1(1984).

11. K. Jain and R. T. Kerth, "Excimer laser projection lithography," Appl. Opt., Vol. 23, 648(1984).

12. S. Rice and K. Jain, "Direct high-resolution excimer laser photoetching," App. Phys. A, Vol. 33, 195 (1984).

13. M. Latta and K. Jain, "Beam Intensity Uniformization by Mirror Folding," Opt. Comm., Vol. 49, 435 (1984).

14. J. R. Sheats, "Intensity-dependent photobleaching in thin polymer films by excimer lasers: Lithographic implications," Appl. Phys. Lett., Vol. 44, 1016(1984).

15. K. J. Polasko et al., "Deep UV exposure of $Ag_2Se/GeSe_2$ utilizing an excimer laser," IEEE Electron Dev. Lett., Vol. EDL-5, 24(1984).

16. D. A. Markle, "The future and potential of optical scanning systems," Solid State Tech., Vol. 27, No. 9, p. 159, Sept. 1984.

17. M. Latta, R. Moore, S. Rice and K. Jain, "Excimer laser projection photoetching," J. Appl. Phys., Vol. 56, 586(1984).

18. D. J. Erlich, J. Y. Tsao, and C. O. Bozler, "Submicrometer patterning by projected excimer-laser-beam induced chemistry," J. Vac. Sci. Tech. B, Vol. 3, 1 (1985).

19. K. J. Polasko, R. F. W. Pease, E. E. Marinero and M. R. Cagan, "Excimer laser exposure of $Ag_2Se/GeSe_2$: high contrast effects," J. Vac. Sci. Tech. B, Vol. 3, 319(1985).

20. M. Nakase, "The potential of optical lithography," Proc. SPIE, Vol. 537, 160(1985); see also M. Nakase,

"Potential of optical lithography," Opt. Eng., Vol. 26, 319(1987).

21. J. Bendig et al., "About the sensitivity of the photoresist AZ 2400 to lasers," Z. Chem., Vol. 25, 158 (1985).

22. E. Cullmann, "Excimer laser applications in contact printing," Semicond. Int., May 1985, p. 332; and Proc. Int. Conf. Microcircuit Engineering, Berlin, W. Germany, 1984.

23. D. Henderson, J. C. White, H. G. Craighead and I. Adesida, "Self-developing photoresist using a vacuum ultraviolet F_2 excimer laser exposure," Appl. Phys. Lett., Vol. 46, 900(1985).

24. P. Dyer and J. Sidhu, "Excimer laser projection micromachined free-standing polymer films," Optics and Lasers in Engineering, Vol. 6, 67(1985).

25. A. L. Bogdanov et al., "Nanosecond single pulse laser lithography," Sov. Tech. Phys. Lett., Vol. 11, 425(1985).

26. V. Pol et al., "Excimer laser-based lithography: a deep ultraviolet wafer stepper," Proc. SPIE, Vol. 633, 6(1986).

27. K. J. Orvek, S. R. Palmer, C. M. Garza and G. E. Fuller, "Resists for use in 248 nm excimer laser lithography," Proc. SPIE, Vol. 631, 83(1986).

28. F. N. Goodall, R. A. Moody and W. T. Welford, "Reduction photolithography by ablation at wavelength 193 nm," Opt. Comm., Vol. 57, 227(1986).

29. R. T. Kerth, K. Jain and M. R. Latta, "Excimer laser projection lithography on a full-field scanning projection system," IEEE Electron Device Lett., Vol. EDL-7, 299(1986).

30. P. E. Dyer and J. Sidhu, "Direct-etching studies

of polymer films using a 157-nm F_2 laser," J. Opt. Soc. Am. B, Vol. 3, 792(1986).

31. G. M. Davis and M. C. Gower, "Excimer laser lithography: intensity-dependent resist damage," IEEE Electron Device Lett., Vol. EDL-7, 543(1986).

32. R. A. Lawes, "Use of excimer lasers in photolithography," Semicond. Int., Vol. 9, 76(July 1986).

33. F. Goodall, R. A. Lawes and P. H. Sharp, "Excimer lasers as deep UV sources for photolithographic system," Microelectronic Engineering, Vol. 5, 445 (1986).

34. B. Hafner and U. Boettiger, "Patterngenerator mit excimer laser," presented at the Ver. Deutsch. Eng. conf. Masktechnik f. Mikroelektronik-Bausteine, Munich, Nov. 1986.

35. J. H. Bruning and M. C. King, "The potential of UV, deep UV and excimer stepper lithography," Proc. Semicon-Japan, 1986, p. C-1-1.

36. J. H. Bennewitz et al., "Excimer laser-based lithography for 0.5 micron device technology," Proc. IEDM, Los Angeles, Dec. 1986, p. 312.

37. M. Sasago et al., "Half-micron photolithography using a KrF excimer laser stepper," Proc. IEDM, Los Angeles, Dec. 1986, p.316.

38. K. Jain, "Advances in excimer laser lithography," Proc. SPIE, Vol. 710, 35(1986); see also K. Jain, "Advances in excimer laser lithography," Proc. SPIE, Vol. 774, 115(1987).

39. T. M. Wolf et al., "The evaluation of positive acting resists for lithography at 248 nm," J. Vac. Sci. Tech. B, Vol. 5, 396(1987).

40. T. E. Jewell et al., "Effects of laser characteristics on the performance of a deep UV projection

system," Proc. SPIE, Vol. 774, 124(1987).

41. M. Rothschild and D. J. Ehrlich, "Attainment of 0.13-μm lines and spaces by excimer-laser projection lithography in 'diamond-like' carbon resist," J. Vac. Sci. Tech. B, Vol. 5, 389(1987).

42. K. Walsh, M. Dunn and J. Bruning, "Performance evaluation of a practical 248 nm wafer stepper," Proc. SPIE, Vol. 774, 155(1987).

43. K. Ogawa et al., "Langmuir-Blodgett films for half-micron lithography using KrF excimer laser and x ray," Proc. SPIE, Vol. 771, 39(1987).

44. A. L. Bogdanov et al., "Computer simulation of the percolation development and pattern formation in pulsed laser exposed positive photoresists," Proc. SPIE, Vol. 771, 167(1987).

45. M. Endo et al., "Half-micron KrF excimer stepper lithography with new resist and water-soluble contrast enhanced materials," Proc. SPIE, Vol. 774, 138(1987).

46. M. Kameyama and K. Ushida, "Excimer laser stepper for submicron lithography," Proc. SPIE, Vol. 774, 147 (1987); see also M. Kameyama and K. Ushida, "The way to one-half micrometer photolithography," Opt. Eng., Vol. 26, 304(1987).

47. M. Nakase et al., "Submicron optical lithography using a KrF excimer laser projection exposure system," Proc. SPIE, Vol. 773, 226(1987).

48. K. J. Orvek, W. C. Cunnigham, Jr. and J. C. McFarland, "An organosilicon photoresist for use in excimer laser lithography," Proc. IEDM, Washington, D.C., Dec. 1987; see also J. C. McFarland, K. J. Orvek and G. A. Ditmer, "Evaluation of an organosilicon photoresist for excimer laser lithography," Proc. SPIE, Vol. 920, 162(1988).

49. I. Higashikawa et al., "Recent progress in excimer

laser lithography," Proc. MRS Symp. 101, 3 (1988).

50. S. G. Olson and M. C. King, "Routes to 0.5 μm lithography," Proc. SPIE, Vol. 922, 300(1988).

51. K. F. Walsh, P. Tompkins and M. D. Dunn, "Performance of a KrF excimer laser stepper," Proc. SPIE, Vol. 922, 396(1988).

52. J. H. Bruning and W. G. Oldham, "A compact optical imaging system for resist process and lithography research," Proc. SPIE, Vol. 922, 471(1988).

53. D. H. Tracy and F. Y. Wu, "Exposure dose control techniques for excimer laser lithography," Proc. SPIE, Vol. 922, 437(1988).

54. D. J. Elliott and J. C. Morgan, "Submicron lithography at 248 nm and 193 nm excimer laser wavelengths," Proc. SPIE, Vol. 922, 476(1988).

55. Y. Ozaki, K. Takamoto and A. Yoshikawa, "Effect of temporal and spatial coherence of light source on patterning characteristics in KrF excimer laser lithography," Proc. SPIE, Vol. 922, 444(1988).

56. M. Takeda and T. Tsumori, "New portable conformable mask method for excimer laser lithography," Proc. SPIE, Vol. 922, 222(1988).

57. H. Nakagawa et al., "An advanced KrF excimer laser stepper for production of 16M DRAMs," Proc. SPIE, Vol. 922, 400(1988).

58. K. Kajiyama et al., "Combination of narrow bandwidth excimer laser and monochromatic reduction projection lens," Proc. SPIE, Vol. 922, 426(1988).

59. F. N. Goodall and R. A. Lawes, "Excimer laser photolithography with 1:1 Wynne-Dyson optics," Proc. SPIE, Vol. 922, 410(1988); see also F. Goodall, R. Lawes and G. Phillipps, Proc. KTI Microelectronics Seminar, San Diego, Nov. 1987, p. 89.

60. B. F. Hafner, "Optical pattern generator using excimer laser," Proc. SPIE, Vol. 922, 417(1988).

61. T. Arikado and T. Takigawa, "Resist heating in excimer laser lithography," J. Appl. Phys., Vol. 63, 1235(1988).

62. M. Endo et al., "New portable conformable masking excimer laser lithography using water-soluble contrast enhanced material," J. Vac. Sci. Tech. B, Vol. 6, 87 (1988).

63. R. L. Woods et al., "Practical half-micron lithography with a 10X KrF excimer laser stepper," Proc. KTI Microelectronics Seminar, San Diego, Nov. 1988, p. 341.

64. C. A. Spence et al. "Deep-UV photolithography with a small-field 0.6 N.A. 'microstepper'," Proc. SPIE, Vol. 1088, 471(1989).

65. J. W. Thackeray et al., "Deep UV AHR photoresists for 248.4 nm excimer laser photolithography," Proc. SPIE, Vol. 1086, 34(1989).

66. K. J. Orvek, C. M. Garza and R. R. Doering, unpublished results.

67. Y. Shacham-Diamond, W. N. Partlo and W. G. Oldham, "Characterization of a UV resist for 248 nm lithography," Proc. SPIE, Vol. 1086, 502(1989).

68. A. Tanimoto et al., "Excimer laser stepper for sub-half micron lithography," Proc. SPIE, Vol. 1088, 434(1989).

69. A. Tokui et al., "Image reversal process using PR1024MB photoresist by KrF excimer laser lithography," Proc. SPIE, Vol. 1088, 462(1989).

70. Y. Tanaka et al., "Sub-half micron lithography with excimer laser," Proc. SPIE, Vol. 1088, 483(1989).

71. Y. Tani et al., "New positive resist for KrF

excimer laser lithography," Proc. SPIE, Vol. 1086, 22(1989).

72. Y. Kawai et al., "Sub-half micron patterning characteristics of silican-based positive (SPP) and negative (SNP) resists in KrF excimer laser litho-graphy," Proc. SPIE, Vol. 1086, 173(1989).

73. A. Ishikawa et al., "Excimer laser exposure characteristics of inorganic resists based on peroxo-polytungstic acids," Proc. SPIE, Vol. 1086, 180(1989).

74. Y. Ichihara et al., "Illumination system of an excimer laser stepper," Proc. SPIE, Vol. 1138 (1989) (in press).

75. J. Wangler and J. Liegel, "Design principles for an illumination system using an excimer laser as a light source," Proc. SPIE, Vol. 1138 (1989) (in press).

76. E. Cullman, "Excimer lasers for lithography appli-cations," Proc. SPIE, Vol. 1138 (1989) (in press).

77. See, for example, J. B. Marling, "Ultraviolet ion laser performance and spectroscopy - Part I: New strong noble gas transitions below 2500 Å," IEEE J. Quantum Electron., Vol. QE-11, 822(1975), and Ref. 78.

78. G. A. Massey, B. P. Plummer and J. C. Johnson, "A high repetition rate ion laser spanning the 195-225 nm spectral region," IEEE J. Quantum Electron., Vol. QE-14, 673(1978).

79. See, for example, J. R. McNeil et al., "Ultra-violet ion lasers," in High-Power Lasers and Appli-cations, K. L. Kompa and H. Walther, Eds., Springer-Verlag, New York, 1979, p. 89, and Ref. 79.

80. K. Jain, "High power, low threshold, cw hollow cathode metal ion lasers," Proc. Int. Conf. Lasers '79, Orlando, Florida, Dec. 1979, p. 442.

81. See, for example, J. J. Ewing, "Rare-gas halide

lasers," Phys. Today, Vol. 31, May 1978, p. 32, and Ref. 82.

82. <u>Excimer Lasers</u>, 2nd ed., C. K. Rhodes, Ed., Springer-Verlag, New York, 1984.

83. See, for example, E. R. Ault, "Table-top Ar-N laser," Appl. Phys. Lett., Vol. 26, 619(1975).

84. See, for example, R. H. Pantell and H. E. Puthoff, <u>Fundamentals of Quantum Electronics</u>, Wiley, New York, 1969, pp. 145-150, and Ref. 84.

85. A. Yariv, <u>Quantum Electronics</u>, 2nd ed., Wiley, New York, 1975, pp. 407-436.

86. T. R. Loree et al., "New lines in the UV: SRS of excimer laser wavelengths," IEEE J. Quantum Electron., Vol. QE-15, 337(1979).

87. T. Takaba, N. Eguchi and M. Takahara, "A new ceramic board exposure sytem using laser-beam and linear motors," NEC Res. Develop., No. 41, April 1976, p. 8.

88. <u>Laser Focus</u>, July 1981, p. 20.

89. R. A. Becker, B. L. Sopori and W. S. C. Chang, "Focused laser lithographic system," Appl. Opt., Vol. 17, 1069(1978).

90. K. Biedermann and O. Holmgren, "Large-size distortion-free computer-generated holograms in photo-resist," Appl. Opt., Vol. 16, 2014(1977), and Refs. 25 and 26 therein.

91. P. A. Warkentin and J. A. Schoeffel, "Scanning laser technology applied to high-speed reticle writing," Proc. SPIE, Vol. 633, 286(1986).

92. M. Lacombat et al., "Laser projection printing," Solis State Tech., Vol. 23, 115(Aug. 1980).

93. A. Kozma and C. R. Christensen, "Effects of speckle on resolution," J. Opt. Soc. Am., Vol. 66, 1257(1976).

94. ✓M. D. Levenson, "High-resolution imaging by wavefront conjugation," Opt. Lett., Vol. 5, 182(1980); see also M. D. Levenson and K. Chiang, "Image projection with nonlinear optics," IBM J. Res. Develop., Vol. 26, 160(1982); and M. D. Levenson, "Photolithography experiments using forced Rayleigh scattering," IBM Res. Report RJ 3770, Jan. 27, 1983.

95. A. Offner, "New concepts in projection mask aligners," Opt. Eng., Vol. 14, 130(1975).

96. L. P. Hayes, K. Jain and R. T. Kerth, "Optical system with diffuser for transformation of a collimated beam into a self-luminous arc with required curvature and numerical aperture," U.S. Patent 4,521,087 (1985).

97. K. Jain and M. R. Latta, "Apparatus for transformation of a collimated beam into a source of required shape and numerical aperture," U.S. Patent 4,516,832 (1985).

98. M. R. Latta and K. Jain, "Optical illumination system using a lens array and a fiber bundle combination," Opt. Soc. Am. Annual Meeting, San Diego, Nov. 2, 1984, Conf. Digest, p. 120.

99. K. Jain, M. R. Latta and G. T. Sincerbox, "Holographic method and apparatus for transformation of a light beam into a line source of required curvature and finite numerical aperture," U.S. Patent 4,444,456 (1984).

100. J. Dyson, "Unit magnification optical system without Seidel aberrations," J. Opt. Soc. Am., Vol. 49, 713(1959).

101. C. G. Wynne, "A unit power telescope for projection copying," in Optical Instruments and Techniques, J. H. Dickson, ed., Oriel Press, Newcastle upon Tyne,

UK, 1969, p.429.

102. S. K. Yao, "Optical consideration in target alignment using nonactinic wavelength microscope," Proc. SPIE, Vol. 772, 118(1987).

103. M. A. van den Brink et al., "Performance of a wafer stepper with automatic intra-die registration correction," Proc. SPIE, Vol. 772, 100(1987).

104. W. R. Trutna and M. Chen, "An improved alignment system for wafer steppers," Proc. SPIE, Vol. 470, 62 (1984).

105. S. Murakami et al., "Laser step alignment for a wafer stepper," Proc. SPIE, Vol. 538, 9(1985).

106. D. R. Beaulieu and P. P. Hellebrekers, "Dark field technology - a practical approach to local alignment," Proc. SPIE, Vol. 772, 142(1987).

107. A. Suzuki, "Double telecentric wafer stepper using laser scanning method," Proc. SPIE, Vol. 538, 2(1985); see also A. Suzuki, R. Hirose and Y. Hirabayashi, "Toward submicron: a new phase of optical stepper," Proc. SPIE, Vol. 632, 166(1986).

108. A. Suzuki, "Laser scanning autoalignment in projection system," Proc. SPIE, Vol. 275, 35(1981).

109. S. Slonaker et al. "Enhanced global alignment for production optical lithography," Proc. SPIE, Vol. 922, 73(1988).

110. J. Goldhar and J. R. Murray, "Injection-locked narrow band KrF discharge laser using an unstable resonator cavity," Opt. Lett., Vol. 1, 199(1977).

111. K. Jain and M. R. Latta, "Beam-Folding Wedge Tunnel," U.S. Patent 4,547,044 (1986).

112. Y. Kawamura et al., "Deep UV lithography by using excimer lasers (photo-etching characteristics

and development of uniform intensity irradiation system)," Tech. Digest of Topical Meeting on Excimer Lasers, Incline Village, Nevada, Jan. 10-12, 1983.

113. J. Pacansky and J. R. Lyerla, "Photochemical decomposition mechanisms for AZ-type photoresists," IBM J. Res. Develop., Vol. 23, 42(1979).

114. C. G. Willson, "Organic resist materials - theory and chemistry," in Introduction to Microlithography, ACS Symposium Series 219, L. F. Thomson, C. G. Willson and M. J. Bowden, Eds., American Chemical Society, Washington, DC, 1983, p. 87.

115. C. G. Willson et al., Polymer Eng. Sci., Vol. 23, 1004(1983).

116. R. D. Miller et al., "Polysilanes: solution photochemistry and deep UV lithography," in Polymers in Electronics, ACS Symposium Series 242, T. Davidson, Ed., American Chemical Society, Washington, DC, 1984, p. 25.

117. Y. Onishi et al., "Polysiloxane with a phenol moiety for bilayer photoresist applications," Proc. SPIE, Vol. 1086, 162(1989).

118. B. J. Lin, "Portable conformal mask - a hybrid near-ultraviolet and deep-ultraviolet patterning technique," Proc. SPIE, Vol. 174, 114(1979).

119. F. Coopmans and B. Roland, "DESIRE: a novel dry developed resist system," Proc. SPIE, Vol. 631, 34 (1986).

120. F. Coopmans and B. Roland, "Enhanced performance of optical lithography using the DESIRE process," Proc. SPIE, Vol. 633, 262(1986).

121. B. Roland et al., "The mechanism of the DESIRE process," Proc. SPIE, Vol. 771, 69(1987).

122. R.-J. Vissen et al., "Mechanism and kinetics of

silylation of resist layers from the gas phase," Proc. SPIE, Vol. 771, 111(1987).

123. F. H. Dill et al., "Characterization of positive photoresist," IEEE Trans. Electron Devices, Vol. ED-22, 445(1975).

124. J. Albers and D. B. Novotny, "Intensity dependence of photochemical reaction rates for photoresists," J. Electrochem. Soc., Vol. 127, 1400(1980).

125. T. Abe, T. Arikado, and T. Takigawa, "Resist heating in excimer laser lithography," J. Appl. Phys., Vol. 63, 1235(1988).

126. N. J. Turro et al., "Organic photochemistry: the laser vs. the lamp. The behavior of diphenyl carbene generated at high light intensities," J. Amer. Chem. Soc., Vol. 102, 5127(1980).

127. R. Srinivasan and V. Mayne-Banton, "Self-developing photoetching of poly(ethylene terephthalate) films by far-ultraviolet excimer laser radiation," Appl. Phys. Lett., Vol. 41, 576(1982).

128. R. Srinivasan and W. J. Leigh, "Ablative photo-decomposition: action of far-ultraviolet (193 nm) laser radiation on poly(ethylene terephthalate) films," J. Am. Chem. Soc., Vol. 104, 6784(1982).

129. R. Srinivasan, "Kinetics of the ablative photo-decomposition of organic polymers in the far-ultra-violet (193 nm)," J. Vac. Sci. Tech. B, Vol. 4, 923 (1983).

130. R. Srinivasan and B. Braren, "Ablative photo-decomposition of polymer films by pulsed far-ultra-violet (193 nm) laser radiation: dependence of etch depth on experimental conditions," J. Polymer Sci., Vol. 22, 2601(1984).

131. J. E. Andrew et al., "Direct etching of polymeric materials using a XeCl laser," Appl. Phys. Lett., Vol.

43, 717(1983).

132. P. E. Dyer and J. Sidhu, "Excimer laser ablation
and thermal coupling to polymer films," J. Appl. Phys.,
Vol. 57, 1420(1985).

133. V. Srinivasan, M. A. Smrtic and S. V. Babu,
"Excimer laser etching of polymers," J. Appl. Phys.,
Vol. 59, 3861(1986).

134. J. E. Andrew et al., "Metal film removal and
patterning using a XeCl excimer laser," Appl. Phys.
Lett., Vol. 43, 1076(1983).

135. M. Rothschild, C. Arnone and D. J. Ehrlich,
"Excimer laser etching of diamond and hard carbon films
by direct writing and optical projection," J. Vac. Sci.
Tech. B, Vol. 4, 310(1986).

136. J. G. Eden, "Photochemical processing of semiconductors: new applications for visible and ultraviolet lasers," IEEE Circuits and Devices Mag., Jan.
1986, p. 18.

137. P. D. Brewer, D. McClure and R. M. Osgood, Jr.,
"Excimer laser projection etching of GaAs," Appl. Phys.
Lett., Vol. 49, 803(1986).

138. M. Hirose, S. Yokohama and Y. Yamakage, "Characterization of photochemical processing," J. Vac. Sci.
Tech. B, Vol. 3, 1445(1985).

139. R. Solanki, W. H. Ritchie and G. J. Collins,
"Photo deposition of aluminum oxide and aluminum thin
films," Appl. Phys. Lett., Vol. 43, 454(1983).

140. G. S. Higashi and L. J. Rothberg, "Investigation
of the surface photochemical basis for metal film
nucleation in laser chemical vapor deposition," Appl.
Phys. Lett., Vol. 47, 1288(1985).

141. P. K. Boyer et al., Proc. Mat. Res. Soc. Symp.,
Vol. 17, 119(1983).

142. K. Suzuki, D. Lubben and J. E. Greene, "Laser-assisted chemical vapor deposition of Si: low temperature (<600 °C) growth of epitaxial and polycrystalline layers," J. Appl. Phys., Vol. 58, 979(19840.

143. R. W. Andretta et al., "Low temperature growth of polycrystalline Si and Ge films by ultraviolet laser photodissociation of silane and germane," Appl. Phys. Lett., Vol. 40, 183(1982).

144. P. K. Boyer et al., "Laser-induced chemical vapor deposition of SiO_2," Appl. Phys. Lett., Vol. 40, 716 (1982).

145. T. F. Deutsch and D. D. Rathman, "Comparison of laser-initiated and thermal chemical vapor deposition of tungsten films," Appl. Phys. Lett., Vol. 45, 623 (1984).

Index

About the Author

Kanti Jain received a Ph.D. in solid state physics from the University of Illinois, Urbana-Champaign, in 1975. Following two years each at M.I.T. and Hewlett-Packard Laboratories, he joined IBM where, from 1979-1988, he was involved in microelectronics and optical technologies programs and held management positions at the Almaden Research Center and Thomas J. Watson Research Center. During 1986-1988 he served with IBM's Corporate Technical Committee as a staff member and senior scientist. In 1988 he founded Anvik Corporation to develop advanced microlithography systems. Dr. Jain is currently Director of Technology Development at Raychem Advanced Packaging Systems. In addition to his pioneering work on excimer laser lithography, Jain's research has focused on deep-UV lasers and applications, nonlinear optical materials and techniques, light scattering in solids, and thin films. He is a Senior Member of IEEE and a member of APS, OSA, and SPIE.